HUMAN NATURE
and the
COSMOS

PAUL BRYAN

Copyright © 2025 *Paul Bryan*

Published by CaryPress International Books
www.CaryPress.com

All rights reserved. No part of this publication may be reproduced, distributed, or transmitted in any form or by any means, including photocopying, recording, or other electronic or mechanical methods, without the prior written permission of the publisher, except in the case of brief quotations embodied in critical reviews and certain other non-commercial uses permitted by copyright law.

Forward

What will become of us? We are not rational. We are a war-prone species of ape with nuclear weapons. We have created such a large population that it is driving a major extinction event. With the now widespread availability of birth control, industrial nations have birth rates that will not sustain their populations. The way we live and feed ourselves has disrupted biospheric processes to the point of sponsoring major climatic change. With females currently ending male hegemony, the genders are separating. We are on the verge of manufacturing artificial items with greater general intelligence than our own.

Considerable evidence exists that extraterrestrials are observing our planet. It may be that they are simply interested in any civilization, or it may be that we are at a particularly concerning place in our development. Perhaps it is not unusual for civilizations to deal with the problems that we are dealing with. Perhaps some survive and some do not. My guess is that civilization on this planet is undergoing increasingly rapid, dramatic, and out-of-control changes, the results of which are beyond comprehension.

This work is an attempt to delineate aspects of what is driving change, and then give some consideration as to what that change may bring. Perhaps you have ideas concerning such things. My hope is that the ideas presented here will inspire future work furthering the understanding of human nature, our place in the cosmos, and where we might go from here.

Contents

Part I – Human Nature 1

Chapter 1 – Religion: A Voice from the Past 3

Chapter 2 – Science: A Pursuit of Knowledge 39

Chapter 3 – Religion or Science: Who Knows? 51

Chapter 4 – Cooperation: An Evolutionary Theme 57

Chapter 5 – Origins: The Ape Brain Rules 89

Chapter 6 – Food: How Our Species Feeds Itself 119

Part II – The Cosmos 149

Chapter 7 – Bottleneck: A Difficult-to-Transition Obstruction 151

Chapter 8 – Imperatives: The Same Everywhere 173

Chapter 9 – Crap: Space Is Not Empty 199

Part III – The Unknown 217

Chapter 10 – People Land: Our Future 219

Bibliography 235

Postscript 247

HUMAN NATURE and the COSMOS

PART I
Human Nature

CHAPTER 1

Religion
A Voice from the Past

Societies change. The Abrahamic religions do not. Faith is demanded of supplicants. If it were to be recognized that one aspect of the canon is flawed, it opens up the possibility of other defects in the religious concrete. Should one error be admitted, the demand for unquestioning faith loses traction.

This conundrum requires the total absence of canonical change. All aspects of religious belief must remain fixed. When society changes regarding what is acceptable and what is not, newly unacceptable aspects of Biblical doctrine are ignored and become invisible. The received Word of God never suffers deletions. Aspects of canons that become unacceptable due to societal change remain in place, but their visibility blinks off like a twinkle light.

The Christian Bible has entries supporting slavery. The Old Testament presents guidelines for owning slaves (Exodus 21:2-27/Leviticus 25:4444-46). Guidelines demonstrate acceptance. When slavery was an ongoing institution in the United States, slave owners cited Ephesians 6:5 and Titus 2:9 (both found in the modernity of the New Testament), in which slaves are admonished to obey their masters, as proof that slavery was OK. Today, slavery is considered totally unacceptable, with Biblical verses demonstrating an acceptance of slavery having become invisible. Ask Christians if the Bible contains anything in support of slavery, and they will most likely deny it and be offended that this was even suggested.

Biblical writing very clearly condemns homosexuality (Leviticus 18:22) along with gender practices, in general, other than the ideals of the ancient society from which the Abrahamic religions emerged. In a startlingly rapid change of mores, homosexuality is currently acquiring greater acceptance. This is no problem for religion. When acceptance of homosexuality becomes as widespread as the rejection of slavery, Biblical condemnations of homosexuality will no longer be noticed. Those denouncing homosexuality by citing Biblical condemnation do not typically mention the condemnation of haircuts found slightly further on in Leviticus 19:27. Maybe God is serious about homosexuality, but just kidding about the haircuts.

The problem is that twinkle lights do not disappear. They just blink off and may blink back on again. Because foundational documents are never deemed to be mistaken and are therefore never erased, they remain available for resurrection by those seeking a return to religious foundations. Pastor Kevin Swanson a Protestant Christian and head of a mega-church, based on information found in the Bible, believes all homosexuals who will not repent and convert to heterosexuality should be killed. He is as closely in touch with his Abrahamic roots as any member of the Taliban and backs up his belief by referring to Leviticus 18:22 and 20:13 in the Holy Bible.

Pastor Swanson, who is as enthusiastic about the Word of God as any Islamic religious leader, unfortunately finds himself on a shorter leash and is unable to immediately act on the sacred knowledge he has acquired. He is waiting for the American Nation to embrace the one true religion (his version of Christianity, just as the leaders of ISIS attempted to establish an Islamic caliphate based on theirs), after which Pastor Swanson and his devotees can get about the business of slaughtering all these gay infidels.

Christian fundamentalists, such as Swanson, who share religious views with Jewish and Muslim extremists, are not as rare as one might hope. Some have attained positions of power in government and the legal system. Roy S. Moore, a former Chief Justice of the Alabama Supreme Court, has described homosexuality as an "inherent evil." Justice Moore lost his position of power attempting to enforce his version of Christian Law in the manner of an Iranian Ayatollah enforcing Sharia Law. He was removed from his judicial position for having instructed Alabama's probate judges not to issue marriage licenses to homosexuals in defiance of a United States Supreme Court ruling widely seen as legitimizing these relationships. Moore's attempt to institute his personal version of the law, rather than the national version, was based on his belief that gay marriage will lead to "destruction of the foundation of our country." Judaism, Christianity, and Islam share a common foundation. The fundamentalists and true believers of all three are in substantial agreement.

If groups such as the Taliban and ISIS were permitted to range free, they would stone, or behead those lacking adequate religious fervor. Additionally, individuals blessed with Godly knowledge may take matters into their own hands. Freelancers attacked the offices of the satirical newspaper *Charlie Hebdo*, killing eleven people whom they considered guilty of blasphemy. Leviticus 24:16, in the Old Testament of the Christian Bible and the source of Islamic beliefs concerning blasphemy, informs us that blasphemers should be stoned to death. I suppose the attackers can be forgiven for going modern and shooting rather than stoning these offenders of the Lord.

Profoundly religious Bangladeshi citizens, as reported in the April 30, 2015, issue of *The Economist*, hacked two insufficiently religious men to death with meat cleavers and murdered an additional two during that month. Also, on a totally individualistic level of responsibility, North African Muslim men, who came to Germany as refugees, took it upon themselves to do God's work and stoned two transgendered women – fortunately they were incompetent, and the women lived. With reference to a profound difference between religion and science, I do not accept String Theory, and yet, to the best of my knowledge, I am not in danger of being murdered by String Theory enlightened physicists.

In some societies, male family members consider it their duty to kill female members who have embarrassed the family by committing a moral transgression with reference to the men's interpretation of Islamic principles. These acts are considered "honor killings" as they are motivated and justified as a means of restoring the family honor that has been damaged by the woman's inappropriate behavior. The 6th of February 2021 issue of *The Economist* discusses women being murdered by family members for such things as their employment, wanting to marry outside their clan, having a Snapchat account, wearing makeup, and so forth. *The Economist* states that "...no country in the Middle East and North Africa releases an official count" of these killings and that "male-dominated governments are doing little to end the killing." And so it goes, on and on.

The almost daily occurrence of deaths by stoning, shooting, decapitation, and hanging sets the Muslim World apart and raises the contention that Islam is somehow intrinsically more violent than Judaism or Christianity. All three have the same Abrahamic roots. All three have fundamentalist adherents. The greater level of violence in the Islamic World derives from a higher percentage of its believers having less secular scientific education, being closer to the precepts of their Jewish/Christian/Islamist roots, having support of a coterie of national governments, and often on longer leashes than their Jewish and Christian fundamentalist counterparts. When Muhammad was inventing Islam, he looked to the Jewish Bible/Old Testament of the Christian Bible as prophet-sponsored sources of the Word of God.

The difficulty lies with foundational Abrahamic ideas not being amenable to change because, as previously stated, if one item is deemed not to be true, it raises the possibility that others may also be out of touch with reality. With science, this is not a problem, as everything is subject to verification and the possibility of rejection. This is a problem for religion because religion demands blind faith in a total package deemed to have originated with God. Science might be a fragmented mess, but religion must remain unflawed concrete because of its demand for faith.

Religious beliefs foster ideas about human nature and how society should be managed. Ideas drive actions. A majority of United States citizens are adherents of an Abrahamic faith. Pew Research Center polls indicate approximately 1.9% are Jewish, 71% Christian, and 1% Muslin, for a total Abrahamic population of about 74%. The sequential listing of Judaism, Christianity, and Islam is based solely on the order in which they originated. The Jewish population is relatively stable, the Christian shrinking, and the Islamic growing. Complexity is added by each of the Abrahamic faiths having a plethora of orthodox and reform factions, religions within religions (e.g., Catholic and Protestant within Christianity), denominations, sects, and cults, each of which believes itself to be precisely in tune with the will and desires of God.

A further complication is the degree to which those reporting membership in a faith are fervent in their beliefs and the extent to which they accept the full gamut of the group's canon. This varies from fundamentalists accepting foundational documents and teachings as the inerrant word of God, to those on the other end of a faith continuum thinking maybe some of it is true, but who really knows? And then there are those who just do not give any consideration to religion and may or may not claim allegiance to a particular faith when questioned.

Additionally, even though the change must remain invisible, religious beliefs do change. This is more than just a twinkling light blinking off, but remaining available for those seeking a return to the true faith as it had originally been presented by God. Contemporarily, science is a big deal with astronomy, for example, having served up ideas about the structure of the Cosmos totally distinct from those that had been presented by religion.

Prior to the rise of science, religion had a lock on everything. The above-referenced move by astronomy into religious territory resulted, for one thing, in Galileo's conviction of heresy in 1633. As an indication of the strength of concern for holding onto all aspects of canon ('What is the big deal to religious beliefs about the Earth orbiting the Sun?'), it took the Catholic Church until 1992 – 359 years later – to admit Galileo's conviction was in error.

It was not bad enough that astronomy was serving up competing stories about the Cosmos. Biology and Darwin came along and started talking about life itself. With the publication of *On the Origin of Species* in 1859, he proposed a scientific explanation for how contemporary lifeforms came into existence in contradiction to the story found in Genesis. He went on to publish *The Descent of Man* in 1871 and *The Expression of Emotions in Man and Animals* in 1872. These works made it clear that Darwin was not simply proposing that only other animals had evolved while humans were created by God in His image. Darwin was talking about all lifeforms. We were, according to his theory, descended from an ancestor we shared with that caricature of ourselves, the chimpanzee. Accepting ourselves as the product of natural selection, and throwing in that even emotions and their expression are products of our evolutionary past and not shared with a divine creator was a hard pill to swallow.

Science introduced us to ideas in total contrast to those supplied by religion. While today just about everyone accepts the idea that the struggles of physics to understand the Cosmos are a more realistic approach than studying religious texts, the conflict started by biology and Darwin rages on. When Darwin published his books in the 19th century, virtually everyone in the Western World was a creationist, believing in God as the creator. Pew

Research Center and Gallup polling provide evidence for the dramatic change in beliefs that has occurred since then.

A Gallup poll in 2019 provided an estimate of 40 percent of the U.S. population being creationists, referring to the belief that Cod created everything, including human beings, who were fashioned in God's image. The poll further provided evidence that a similar percentage accepted evolution. A Gallup survey in 2017 had found evidence that 24 percent of the U.S. population accepted Biblical veracity in its totality. We have 74 percent declaring membership in an Abrahamic faith, while somewhere around 40 percent adhere to the concept that God manufactured everything, and something in the neighborhood of 24 percent accept the Bible as the true and precise word of God. This is quite a change from almost total acceptance of the religious canon. Also of great interest is that this change is ongoing.

All polling numbers are, of course, approximations. While Gallup and Pew Research do a competent job of selecting persons to represent a particular population, a serious issue centers on how questions are worded. Variations on the wording of a question can lead to dramatically different results. Pew Research and Gallup put a great deal of effort into question structure, but the presented statistics must be viewed as ballpark figures.

In 2019, Pew Research pursued concerns about question structure by means of exploring attitudes concerning religion and evolution. Respondents were asked their opinions on creationist and Darwinian beliefs in two different ways. One half of the subjects received one form of the question while the other half received the other. In one form, respondents were asked a two-part question, the first part of which was if "humans and other living things had evolved over time" as advanced by Darwin, or if "humans had existed in their present form since the beginning of time" as presented by the creationist view. The second part of the question was then directed to those indicating a belief in evolution, asking whether or not God was involved in the process.

A second version consisted of a single question with a presented choice of three possible answers. The choices were:

1. "Humans have evolved over time due to processes such as natural selection; God or a higher power had no role in this process."

2. "Humans have evolved over time due to processes that were guided or allowed by God or a higher power."

3. "Humans have existed in present form since the beginning of time."

Apparently, with the first version, by the second part of the question, respondents had already selected whether science or the Bible was the primary conveyer of truth. With this question structure, 31 percent chose the creationist position, indicating a belief that evolution had not taken place and that people were the same as when our species had been manufactured by God. With the second presentation of the question, with three possibilities to immediately choose from, only 18 percent chose the creationist view that people were the same as when God first created them.

The first part of the first version appears to have presented respondents with the necessity of choosing either evolution or the Bible. By the second part of the first wording, that decision had already been made. With the alternate wording of the question, evolution can be chosen with commitment to religion remaining in place. While both versions aimed at determining the extent of belief in evolution in relationship to Godly creation, the wording of the question produced a dramatic difference between 31 and 18 percent. Researchers put considerable effort into their work, but the resultant statistics require evaluation with a grain of salt.

Statistics generated by Gallup and Pew, even though of questionable precision, provide a great deal of support for the contention that change in religious beliefs is not only continuing but ramping up. The previously referenced Gallup survey was conducted 160 years after Darwin published his work. At the time of Darwin's publication, belief in Biblical veracity was almost universal in the Western World. The new Gallup data indicated a drop in belief had occurred at about one-third of one percent per year.

Pew Research Center Religious Landscape Studies in 2007 and again in 2014 found persons stating they were Christian dropped from 78.4 to 70.6 percent, and those indicating they were atheist, agnostic, or of no particular faith increased from 16.1 to 22.8 percent. While the change between Darwin's publication in 1859 and Gallup's survey in 2019 amounted to about one-third of one percent per year, the Pew work found a change that had ramped up to around a full percentage point a year. Change was not simply continuing, it was ramping up.

The Pew study further demonstrated differences in religious views between young and old, supporting the contention that the trend of speedier change will continue. Whereas 11 percent of those 70 and older identified themselves as "atheist, agnostic, or nothing in particular," that figure climbed to 36 percent with 18- to 24-year-olds. This is a significant difference in religious intensity. By the time these younger people reach their 70s, they may have become more religious, but Pew Research argues that differences will remain with "generational replacement" driving religious decline.

Contention exists concerning whether or not fewer people reporting to be Christian actually reflects a decline in religiosity, or simply the fact that people are experiencing a realignment of spiritual values reflecting greater individuality. The argument is that people are every bit as religious, but simply in a less conventional manner. Arguing for an actual decrease in religiosity is that of those within the "atheist, agnostic, or no particular faith" group, the percentage of those reporting to be atheist or agnostic went from 25 percent of that category to 31 percent, with a concurrent drop in the "no particular faith" group. It is difficult to justify the assertion that those going from "no particular faith" to "agnostic" or "atheist" is simply a shift toward greater individualism and does not demonstrate a lessening of religiosity.

The just-presented decline in religiosity occurred in conjunction with an increasing acceptance of evolution as a natural process devoid of godly interference. Whereas a 1999 Gallup Poll found 9 percent of respondents purporting to be hard-core secular evolutionists with no role played by God, that percentage had increased to 19 by 2014. Considerable evidence exists that science has been replacing religiosity since Darwin's time and that this process has sped up.

Concern now centers on whether or not science will continue to displace religion. Three ongoing areas of social change argue that it will. First, young people are less religious than older generations were at that same age. Second, education levels are rising, and an inverse correlation exists between education and religion. Third, society is becoming more urban, with an inverse correlation existing between urbanity and religion.

Religiosity varies over the lifecycle, with people having lower levels as young adults and becoming more religious as they grow older. When asked about the importance of religion in their lives, 47 percent of those 18 to 29 years of age reported it as very important, among 30 to 49 years old, this increased to 56 percent. With the 50 to 64 years group, it had risen to 60 percent, and for those 75 and over, it had risen to 75 percent. While levels of religiosity among the young will almost certainly rise as they age, the issue is whether or not they will do so to the point of fully matching those of the older generations or whether at these later ages they will retain lower levels of religiosity reflecting a continuation of the social change that has been proceeding since Darwin.

Pew Research Center, based on a 2010 survey, reported that younger members of the population, as expected, were less religious than older. Those born from 1965 to 1980 reported lower levels of affiliation to an established faith than did those born from 1946 to 1964. Of interest is that the older group reported higher levels than did the younger when they had been at that younger age. Twenty percent of the 1965 to 1980 group reported being unaffiliated with any particular religion, while those in the 1946 to 1964 age group reported only 13 percent as unaffiliated when at that same age. If the sample born more recently reports lower levels of religious affiliation than the sample born before this reported for that same age, it reflects social change. When the younger population ages, their lower levels of religiosity may reasonably be expected to continue.

Additionally, Pew, using 2008 General Social Survey data concerned with whether or not a belief in God was held with absolute certainty, presented data indicating that while 53 percent of 18-to-29-year-olds reported a belief in God with absolute certainty, 55 percent of 30-to-49-year-olds did so at that same age. While a change of only 2 percent may not seem like much, this rate, if accurate and maintained, adds up. A change in beliefs concerning God is profoundly unsettling and may reasonably be expected to proceed slowly. Concerning the belief in creationism, a decline from just about everyone to around half of the population took in the neighborhood of 150 years.

Educational attainment is increasing for citizens of the United States. In 2006, 28 percent had earned a bachelor's degree or greater. By 2012, that had risen to 33 percent. The percentage of those having attained a high school degree and those with a high school degree plus some college had also increased. High school completion went from 86 percent to 90 percent, and high school completion with some college went from 58 percent to 63 percent. Additionally, with more people completing more education, as would be expected, younger people have greater levels of educational attainment than their elders.

The relevance of the preceding educational changes lies in the inverse correlation between education and religiosity. Gallup and Pew both found, with multiple research efforts, that greater education means less faith. This includes such aspects of religion as its importance to the individual, certainty of belief, and frequency of prayer.

The United States is becoming more urban, with a continuing migration from rural to urban areas. In 2000, based on U.S. census data, 79 percent of the population was urban. Ten years later, in 2010, the urban population had climbed to 80.7 percent. Gallup data indicates religiosity is higher among rural residents than among those living in the suburbs or urban areas across all age groups. The question then becomes: will the migrants change the city, or will the city change them? One influential factor is that a high proportion of migrants are young people looking for jobs. Young people are the least religious and the least set in their ways. As they acquire new peer groups, they may typically be expected to change to fit in with the new people and beliefs.

Religion is presented as foundational to morality and decency, without which life would be characterized by bad behavior and rampant selfishness. A major concern is whether or not declining religiosity will result in societal decay: with less religion, will society become less humane and nastier? Science, specifically regarding primate studies, presents a contrasting argument.

Primatologists, such as Frans de Waal in *Chimpanzee Politics: Power and Sex among Apes* and his more recent work *The Bonobo and the Atheist,* argue that not only did bodies and emotions originate in our evolutionary past, but the roots of human morality as well. They contend monkeys and apes have a sense of fairness in which one good deed is seen as deserving of another, and reprisal is appropriate for bad deeds. These feelings of acceptable and unacceptable behavior are held to be an evolved biosocial phenomenon in which genetically sponsored predispositions mesh with socialization, and that this is absolutely essential for the existence of primate societies. Herein lies the roots of the Golden Rule and human morality.

Primate societies (including our own), although unified, are anything but orderly. They are not, however, a war of all against all as Thomas Hobbes would have it for both chimps and people. Referring once again back to Frans de Waal's 1982 work with chimpanzees, we see that even though one individual may be the strongest, it cannot dominate without the support of associates: coalitions form in which pairs or groups dominate through cooperative intimidation of what may very well be a stronger individual.

Further, dominant individuals and cliques are subject to group censure. De Waal refers to several observations of a dominant male, apparently after having behaved in a manner widely considered to be beyond the pale, being chased by a group of females. It is the give and take between displays of power and widely recognized ideas of what is acceptable, or morally permissible, that makes a chaotic primate society capable of teamwork essential for the maintenance of territory and resources.

Social order, or peaceful coexistence, breaks down when monkeys or people are involved in situations in which they perceive themselves to be getting a bad deal. Michael Shermer, in the Skeptic column of the May 2015 issue of *Scientific American*, presents an experiment Frans de Waal conducted with capuchin monkeys in his Emory University laboratory. Two monkeys, in side-by-side cages and clearly visible to each other, are trading rocks for cucumber slices, which they then eat. One of the monkeys is given a grape in exchange for its rock, rather than a cucumber slice. After that monkey receives the more valued grape, the other monkey is, once again, given a cucumber slice. Apparently feeling that this is not fair, it throws the cucumber slice back at the experimenter, rattles its cage, and slaps the floor. Receiving a cucumber slice when someone else has just gotten a grape is a bad deal, leading to agitation and a pitched fit.

De Waal contends morality is an evolved attribute we are predisposed to learn, with his thesis being that our sense of fairness or right and wrong has roots in our pre-human past. For primates, in general, social interaction meshing with developing cognitive capacity produces an adult with a sense of fairness or morality. For humans, the acquisition of morality is similar to the acquisition of language. We learn language through interaction with others when our brain attains an appropriate level of development. The argument goes that our sense of morality is similarly acquired through socialization during a developmental period, particularly amenable to its acquisition.

While chimpanzee morality and sense of fairness incorporates the concept of "one good deed merits another," reciprocity also has its dark side, with revenge being appropriate for bad deeds. Just as de Waal's works are replete with examples of reciprocity in which chimps respond to such things as grooming, greetings, or shows of affection with another good turn such as food sharing or agreeable physical contact, they are also replete with tales of revenge following a bad deed, such as a launched attack involving one chimp knocking down and stomping on another.

It is pleasant to think we do better than apes when it comes to revenge. Unfortunately, most human homicides are of the "one bad turn deserves another" sort and occur in the context of family and other interpersonal relationships. While we like to think of ourselves as being rational, we most commonly kill for revenge. If money is involved, it is frequently not because we are seeking to acquire money through an activity such as armed robbery, but rather because we feel victimized by a bad deed and that revenge is appropriate. We, much like other apes, do most of our intra-group violence because of our emotions and, as incongruent as it seems, our sense of morality with its "one bad turn deserves revenge" component.

An interesting aspect of chimpanzee morality and sense of fairness is that it only applies within the individual's group. Outsiders are beyond the pale and subject to mistreatment regardless of prior or current interactions. Richard Wrangham arrived at Tanzania's Gombe National Park in 1970. Jane

Goodall's chimpanzee studies were underway at this location, and the local chimpanzee community had grown large enough to begin the process of fission into two groups: northern and southern. As the separation progressed, chimps in the newly forming communities became increasingly hostile to one another. Parties of six or so males, with an occasional female (United States combat forces now admit women as full participants and have finally caught up with our chimp cousins), began to patrol the newly established border and make forays into each other's territory.

Should one of these patrols come across an isolated individual from the other group, they would attack, inflicting injuries that were often fatal. Over several years, the southern band was annihilated, and the northern chimps, as part of this process, gained control over territory containing nutritionally consequential fruiting trees. Other studies of chimpanzee interactions in their natural setting demonstrate the normalcy of the forgoing process. Chimp morality is for members only.

If human morality is supplied by God, as opposed to arriving through evolution from our primate progenitors, perhaps it should differ from chimp morality by extending to those beyond our own group. Unfortunately, if chimps could write foreign policy, they would produce what would go hand in glove with that of the ancient Israelites, put forward as God sponsored, in the Old Testament. In Numbers 31:18, Moses (who is in direct communication with God and merely passing on His demands to the Israelites) orders the slaughter of just-conquered Midianites except for the virgin females, who may be raped and kept as slaves. This is chimp foreign policy put into words and justified by religion.

Steven A. LeBlanc, in his 2003 work *Constant Battles: The Myth of the Peaceful, Noble Savage*, argues that archeology provides evidence for the normalcy of violent and fatal conflicts between human groups. He contends that archeological finds have frequently been misinterpreted through an ideological desire to see our past and nature as peaceful, with war held to be the product of historic civilizations.

As an example of the preceding, LeBlanc refers to the initial theories advanced concerning the five-thousand-year-old body known as "Otzi, the Ice Man," found in 1991 thawing out of a glacier in the Alps. First speculation raised the possibility of Otzi being a trader traveling on business, or a shepherd tending a flock, who had gotten caught by changing weather conditions and died in a snowstorm. LeBlanc points out, however, that X-rays revealed an arrowhead in his upper torso, and DNA testing disclosed the blood of four other persons on his equipment and clothing (one on his knife, two more on an arrowhead, and a fourth on his coat). A copper implement initially referenced as a hatchet, evidenced no wear marks indicating it had ever been employed to chop wood, and given the malleable nature of copper, it would have been too soft to have effectively done so – a more realistic classification is that of a battle axe.

Moving on to the historic era and continuing the discussion of the extent to which our morality applies to outsiders, when Europeans came to North America, it was already occupied by civilizations that, unfortunately for them, were militarily inferior. Our morality permitted the slaughter of these outsider people and the building of our country on land taken from them.

More recently, the United States invaded Vietnam, where, to protect our own country from the falling dominoes of communism and save Brown People from the inevitable failure that would surely arise from their attempts to run their own country, we bombed them with a mixture of high explosives, burning napalm (gelatinized gasoline), and poison chemicals (agent orange) leading to the death, mutilation, and birth defects of untold numbers of human beings in our morality-sponsored attempt to save ourselves from a takeover by godless communism and bless them with democracy.

It has been a long time since the appropriation of native lands in North America, and over fifty years since Vietnam. Perhaps a God-sponsored morality has seeped into our relationships with those beyond the American pale. Unfortunately, foreign policy is still "same old, same old." On October 3rd of 2015, a United States AC-130 gunship – apparently purposefully so – destroyed a "Doctors Without Borders" hospital (Mathew Rosenberg, 2016). I say "apparently purposefully so" as the hospital was well lit, its roof clearly marked, and the attack continued for approximately 1 hour and fifteen minutes despite efforts to contact the United States and NATO military to get it stopped. It stopped not with the pleading but when the job was finished. Looks like apes with technology to me.

It is not my intention to single out the United States as especially prone to the preceding sort of behavior. Americans are normal members of our species with normal in-group/out-group behavior. As an example, slightly over a month after the hospital incident, a group out of the Middle East known as ISIS staged multiple coordinated attacks in Paris, France, inflicting a wide assortment of rude and offensive behaviors, numerous casualties, and deaths. Even though there are occasions when we whack our own, this sort of enthusiastic slaughter – with no apparent guilt and a great deal of self-righteous piety – is reserved for outsiders. Just as with chimps, moral precepts apply only to the group in which we have membership.

While chimpanzee social life is more complex than any other nonhuman primate that humans have observed to date, it remains considerably less so than our own. For one thing, chimp communities are homogeneous, and human communities are heterogeneous. In the past, human communities were also homogeneous, but with empires, slavery, colonialism, and industrialization, cultural diversity became the norm. Contemporary communities consist of an assortment of intermingled peoples with widely varying levels of hostility between them – we live with outsiders the likes of which chimps only deal with on their borders.

Much as would be true for chimpanzees, our morality fails to constrain problems that develop between these juxtaposed groups, with results that include lynchings, murders, riots, hate crimes, and a wide assortment of rude and offensive behaviors. Historically, cities in the United States have experienced racial violence along with ethnic rioting between immigrants and more established populations.

Isabel Wilkerson, in her work *Caste: The Origins of Our Discontents*, reports on the "wave of anti-black pogroms in more than a dozen American cities" that took place in conjunction with Black migration north following the Civil War (pg. 229). Much of this violence focused on the businesses and homes of Blacks, evidencing success and prosperity. These were out-group communities that were highly visible because they were successful in ways valued by the larger society. Today, racial tensions continue to exist with new antagonisms between Christian and Islamic groups, and now between political entities.

Distrust, hostility, antagonism, hatred, and attendant violence are justified by those involved through unrealistic assessments of both their own group and those whom they consider to be outsiders. While problem behaviors and faults of the group to which we belong remain largely unperceived, those of outgroups are microscopically examined and exaggerated to the point of considering these people to be inferior, possibly to the point of being less than human. We are God's chosen people, and the outsiders are Tolkien's Orcs.

Given the preceding, and returning to the underlying concern of this work: is human morality godly or apish? The religious set typically presents a perspective of good behavior having arisen with godly input. Unfortunately for this perspective, our morality appears identical to that of other primates. All of us appear possessed by the idea that a good act deserves a pleasant response, while a response to a bad deed includes a revenge component. It is also quite interesting that, across the board, morality only applies to groups in which we are invested. Greater differences between ourselves and other primates could reasonably be expected if human morality had been installed by God, whereas the morality of other primates resulted from evolutionary processes.

The key to understanding why morality extends only to other community members lies with the comprehension that we, as with chimps, cooperate within our communities to compete more effectively with other communities. Morality is a term used to describe rules governing interactions within a community that serve to facilitate coordinated behavior, permitting more effective competition for resources. We exhibit moral behavior not because we are good, decent, and Godly, but because it is an evolved characteristic conferring a survival advantage by way of assisting our group to overpower neighboring groups. As previously mentioned, the Blacks moving north following the Civil War and being financially successful were seen as an alien community unfairly pursuing white values. If morality applied across the board to all communities, Blacks moving north and being successful would have been congratulated, Native Americans would still own America, and it would not matter all that much who won the Super Bowl.

Chimpanzees have one level of community. Loyalties are confined to the individual's immediate group. The collection of chimp groups located within Tanzania's Gombe National Park, for example, they exhibit no apparent loyalties, pride, or patriotism accruing because they are all Gombe Chimps. Competition for chimpanzees remains a simple matter of adjacent groups competing for territory and accompanying resources.

Both loyalties and competition exist on the various levels of multi-tiered human communities. For the modern world, the intergroup violence once practiced between adjacent communities now takes place between groups within nations, between nations, or between nations and other strata of social organization with which they are in contention. Extended family groups and small towns no longer raid each other, a la chimpanzees and preliterate peoples. High school sports teams, however, symbolically carry out the raiding practices, mostly minus the bloodshed, but with the full complement of *strum und drang* generated by brain parts shared with our ape cousins. This makes sense of how it is that the victories and defeats of sports teams can be followed with such enthusiasm.

The communities of advanced social species typically find their main competition to be other communities of their kind: chimpanzees dominate the local fruit supply, with the most consequential competition coming from other chimp groups. With weapons limited to body parts, the neighbors might be wiped out and valuable fruit trees appropriated, but these activities do not threaten species survival. The same cannot be said for our nuclear-armed species. While immense differences exist between chimp tool use and our own, human and chimp morality may have greater similarities than we typically care to recognize.

Going back to a central concern of this work: is human morality godly or apish? Human morality has been shown to contain implicit acceptance of both "good for good" and "bad for bad" interactional responses, and to concern only interactions with other entities within one's perceived group. If God had engineered our sense of "right and wrong," it might reasonably be expected to diverge at least somewhat from that of other primates. A reasonable guess is that we are apes in possession of considerable technology.

Our next concern centers on the value of religion to society. On a very basic level, morality benefits human groups the same as with other primate groups: whatever is seen as beneficial is perceived as reasonable and justified. If fruit trees and the land they occupy are recognized as of value to us, their ownership by outsiders is viewed as wrong. Religion, as an evolved attribute of our species, does not make us nicer or kinder. It conveniently provides further justification for selfish interactions with others. We are now imbued with the emotion-laden belief that God intends or wants us to possess the desired commodity. If it is land desirable for its resources or our occupation, current residents may reasonably be removed by necessary means. Following the seizure and displacement of prior residents, it may then be assumed that God had granted us this land.

Although religion evidences a "three unaware monkeys" position when it comes to the slaughter of inconvenient natives, perhaps it serves to amplify overall levels of graciousness and civility within a society. Perhaps a relationship exists in which higher levels of religiosity are associated with lower levels of bad behavior. Unfortunately, that does not appear to be true.

The Southeastern area of the United States evidences the highest overall levels of religiosity based on Gallup research concerning such things as belief in God and church attendance. In contradiction to the preceding idea that more religiosity leads to less bad behavior, the Federal Bureau of Investigation's Uniform Crime Reports point out that, with both property crimes and crimes against persons, this area has the highest overall levels of bad behavior. It has the most religion and the worst behavior: go figure.

It is not just within the United States that the above relationship appears to hold. Europe has lower levels of religiosity than the United States, along with lower crime rates. Germany, as an example in comparison to the United States, has a smaller percentage identifying as Christian and one-sixth the homicide rate.

While it does appear that a positive relationship exists between religiosity and behavior justifiably considered to be bad, surely those inhabited by the Holy Spirit would be more altruistic than their secular neighbors. Jean Decety, a neuroscientist at the University of Chicago, sought a connection between religiosity and altruism by studying generosity among children, the assumption being that generosity would serve as an indicator of altruism. The research found an inverse correlation between religiosity and generosity. Children from more religious families were less generous with their classmates than children from less religious families. Heightened religiosity appears to intensify a sense of boundaries between groups. The key to making sense of both why heightened levels of religiosity would be associated with heightened levels of bad behavior and why children from more religious families would be less generous than children from less religious families lies with an evolutionary perspective.

Primate morality serves to enhance coordinated activity within a community by limiting conflict through the establishment of standards of what is acceptable and what is unacceptable behavior, a more peaceful community being propitious to coordinated activity. Life in a chimpanzee community is not a Hobbesian war of all against all. As chaotic as it may appear, with individual members vying for power and sex, violent interactions are limited through agreed-upon standards (male chimpanzee conflicts over power and sex are typically not life-threatening) and moderated through routine gestures of dominance and submission (conflicts are settled with recognized symbolic gestures). The result is a community in which groups of its members are capable of presenting a united front in competition with neighbors for territory and resources.

Standards of behavior, or morality, within a community do not apply to the neighbors. If the standards of acceptable and unacceptable behavior within the community were extended to outsiders, the result would be a loss of competitive capabilities. These others would then need to be given some consideration, and we could no longer simply gang up on them and seize whatever we could. Within the community, it sponsors civility. The other side of the morality coin sponsors a view of the neighbors as beyond the pale due to their ignorance concerning supernatural matters and more appropriate for exploitation than altruistic interactions. European conduct when the Americas were being appropriated serves as an excellent example.

The evolution of human religion is a further stage of the sociality theme in which a newly evolved religion melds with already extant primate morality. With religion, a community sees its way of life as being in line with the desires of supernatural forces, and its territory held at their behest. It amalgamates with primate morality in that standards of acceptable and unacceptable become sanctified by powers beyond the natural world. Religious content, or its canon, is presented within a linguistically transmitted story (remember Adam and Eve), figuratively and symbolically telling how their people and the world came into being, how their society conducts itself, and how its members are to behave. Human religion supercharges primate morality by affirming that a community is held to be special by forces beyond the natural world. When in conflict with the neighbors, we are now carrying out the will of a divine power and no longer just jacked up on endocrine secretions shared with chimpanzees. The endocrine secretions are still there, but we now additionally do not want to disappoint or piss off God.

Moses was representing a pissed-off God when he informed the Israelites that they were not just supposed to kick the Midianites' asses and take or destroy everything they owned. God wanted the Midianites, to include the women, slaughtered, with the exception of the virgin females, who should be – how convenient is this? – kept as slaves (Numbers 31: 1-8). If that is not what a first-class patriarchal God would want, I do not know what is. And, by the way, the big reason the Midianites' women deserved to be treated so harshly was because some had seduced Israelite men. The temptresses had caused Israelite men to behave badly in the eyes of the Lord. Even back in Bible Times, guys could not keep their dicks in their pants and it was not their fault.

Just as morality can only apply within the community, supercharged religious morality must also only apply within the community, or the evolutionary value driving its existence is lost. When human communities changed from foraging to farming and coalesced into larger political organizations, religion changed in step with them. The formation of larger political groups required religions of a size specific to each group. Ancient Israel was cemented together with Judaism, and Christianity was convenient for the Roman Empire. As communities of various sizes form or fragment, religions coalesce or shatter to fit them. It is highly unlikely that the splinters of Islamic faith found in the contemporary Middle East, and serving a wide range of cultures, can be successfully fused to confirm the sanctity of a Caliphate founded through ISIS violence.

Through colonialism, immigration, war, and globalization, societies have lost their homogeneity. The intermingling of diverse groups in contemporary, heterogeneous societies means there may not be a shared sense of community with the next-door neighbor or the kid in an adjacent seat at school. Religiosity does not promote a generalized love of mankind. It promotes solidarity within a community and a sense of superiority over neighboring infidels. The neighbor may be a stranger. A heightened personal sense of religiosity only adds to the perception of her or his appropriateness for exploitation. A kid from a religious family has been assured she or he is one of God's chosen people and are destined for eternal life in paradise. Why should he or she be generous with someone who is apparently out of touch with divine powers? Religion exists because it enhances competitive advantage and has done so to the extent that all our non-human hominin relatives were driven to extinction. It is a big "fuck you" if you do not happen to be one of us.

A religious contention is that people need input from God to arrive at moral values. It has been argued that, contrary to this, our morality originated with evolution and has its beginnings with other primates. The morality supplied by the Abrahamic religions has at its base the values of an ancient, patriarchal society. It is necessary to come up with concepts more appropriate for a contemporary society. A god did not supply people with concepts of right and wrong because gods are human inventions and do not actually exist. We have been making this stuff up all along.

Religion, rather than providing universal guidance of what is right and wrong, supports what exists within a group. It provides support by endorsing community values and ways of doing business. The news and views originating from an ancient patriarchal society endorse and back-up the ways of doing business within that society, with its sexism and negative attitudes toward the human body and sexual activity.

Religion also supports contemporary leaders and those in power. Kings typically consider themselves to be serving at the will of God. Those in power latch onto religion to justify and maintain their way of doing business and their personal power. They are sincere in their beliefs. Community social structure and their own dominant place in it just seems right to them.

Foundational Abrahamic concepts and documents justified Abrahamic society and its behaviors. Bible parables presented citizens with a view of social construction and behaviors indorsing societal mores, folkways, and leaders. The ancient Israelites were a male dominated agricultural society. They invented a religion and a God to justify their way of life.

The Abrahamic creation story is presented in the Hebrew Bible and the Old Testament of the Christian Bible and underpins the Islamic Qur'an. In Genesis 1:26 – 2:15 of the Bible, God manufactures Adam in his own image, which includes that He and Adam have some traits in common. A male-dominated society invents a male God: males, purportedly having been created by this God, share Godly attributes. It is all rather flattering for men. Certainly, more so than having our chimp cousins pointed out to us.

It needs to be kept in mind that this creation story was concocted before the invention of primatology. As a rule, our species does not accept blank areas of knowledge. If we do not know or understand something our creative minds fill in the blank. Now, with primatology, we have a more realistic possibility of getting our origin story straight and recognizing that the Abrahamic attempt does not mesh with or fit into reality in any way.

According to the God thing, the male gender was invented and manufactured first. After manufacturing Adam, God decides it might be nice if he had some company and cobbled together Eve. In doing so God started with a body part taken from Adam. Eve was fashioned second. This established male primacy and men as the template for God's creation. In contrast to this, Mattel created Barbie first and Ken as something of an afterthought.

The reality of human biology does not support the religious contention for male primacy. Fetal development supports an argument for a female template. "Females are considered the 'fundamental' sex – that is without much chemical prompting all fertilized eggs would develop into females" (DeSaix and Johnson, 2017). If a fetus is a genetic male, the testes produce androgen hormones that result in a male body structure. If these hormones are not present, or a condition exists in which the fetus lacks sensitivity to them, the fetus develops in a female configuration regardless of genetics. This makes the male an altered version of the female, achieved by means of hormonal influence during fetal development.

Eve behaved badly by eating a fruit that God had forbidden them to eat. In addition to Eve behaving badly herself, she talked Adam into joining her in the snack. This Biblical parable establishes that women are not to be trusted. Adam would not have eaten that fruit if Eve had not talked him into it. The tale establishes that women have bad judgment and are not to be trusted. If men behave badly, the underlying cause is often a woman. Male hegemony is established as a reasonable condition given the nature of women.

Our society has justified denying opportunities for full societal participation by women through beliefs concerning the nature of women that make them unsuited for many roles. Women are, or at least until very recently were considered to be, too emotional, timid, and delicate for full participation. Their place was with children and in the kitchen. Religious beliefs did not create this situation. Religious beliefs reflect it. Religion backs up culture by presenting beliefs as natural and the way God wants things. It is not men's fault that women are denied full societal participation. As the story goes: it is the nature of women that makes this necessary.

Abrahamic foundational documents also sponsor a negative attitude toward the human body. The fruit that Eve inappropriately ate, and then persuaded Adam to join her in the snack, provided them with knowledge of good and evil (Genesis 2:9). Because they now know about good and evil they recognize that they are naked and have invented porn, and they are ashamed and hide from God (Genesis 3:8). God (being God and therefore totally brilliant) immediately guesses what has happened and invents tailoring to do what he can to ameliorate the situation (Genesis 3:21).

If the human body is, in and of itself, shameful, it is a good guess that sex is not to be taken lightly. A woman who is found not to be a virgin at marriage is to be stoned to death (Deuteronomy 22:13-20). When Onan (Genesis 38:9-!0) engages in what could be either coitus interruptus or masturbation, God kills him. Ideas about sex, as found in Biblical writings, create problems for us today. They are also reflective of a patriarchal society: it is women who must remain virginal.

In Paul's letter to the Corinthians (I Corinthians 7:1-2), he recommends marriage as a means of managing sexual desire in a manner that is OK with God. According to Paul, it would be ideal for men to never touch women, but they would be too horny and lacking in self-control to pull it off. Marriage is better than burning in Hell for all eternity, and is, of course, an essential condition for the fabrication of children. God, possibly experiencing mental health issues, made humans sex crazed but with strict instructions as to its expression.

Sex is a genetically programmed biological drive. It cannot be prevented from starting up or turned off at the convenience of social structure. Just as with a mole in the in the Whack-A-Mole game, sex repressed in one area pops up in another. Whack-A-Mole Sexual Theory argues that the denial of a sexual outlet for Catholic priests will have its problems. The extent of these problems was brought to light by a Boston, Massachusetts newspaper, The Boston Globe, in 2002. The Globe, in pursuing stories of priest sexual misconduct in the Boston area, found that about half of them engaged in sexual activity with other adults. The recipients of this priestly ardor apparently varied widely in the extent to which they were happy campers. The most disturbing finding was that an estimated six or seven percent of priests were engaging in sexual activities with children.

The only humans socially deemed appropriate for sexual activity are spouses, and priests are forbidden to marry. They have no acceptable outlet for a biologically driven behavior. Since the 2002 Globe report, evidence of priests sexually engaging with and/or abusing both adults and children continues to emerge. In 2018, a Pennsylvania law enforcement investigation and grand jury report found ongoing sexual abuse of adults and children by Catholic clergy. Additionally, evidence was found of these activities being covered-up by church officials.

Response from within the Church is to express shock and remorse, and provide assurances that solutions are now in place to eliminate the problem. These efforts center on the issuance of directives emphasizing the necessity of sexual purity, and holding supervisory personnel accountable for reporting the unwanted behavior of subordinates. Due to the widespread nature of the behavior, it is almost certainly an open secret with those responsible for reporting, as likely as not, to also be participants.

The Church's "no sex" policy includes "no masturbation." As just about all men masturbate, along with most women, concealment of this forbidden activity must be commonplace. The standard practice of concealment in one area sets the stage for concealment in others. With masturbation already covert, sex with others, to include both adults and children, is added to the category of superficially concealed but actually commonly known behaviors. It is extremely unlikely that the level of activity for both masturbation and sex with others could be a well-kept secret.

Priests are not the only members of the population to lack sexual outlets and suffer harm due to ideas foisted on contemporary society from an ancient civilization. The physically handicapped share sex drives along with the rest of us. Ideas concerning human nature and sexuality, presented as God given, lead to the unavailability of sex for these people.

A number of years ago, the Dear Abby column dealt with the sexual needs of physically handicapped persons. A friend of someone who was paralyzed and confined to a bed and a ventilator in a room at his parents' home arranged with an escort service for him to have a sexual experience. His parents discovered the incident and cut off his contact with the person who had arranged the encounter. They strongly approved with ideas propagated by the invented God and religion of a three thousand or so year old civilization.

The column generated a great deal of widely varying responses. Those confined to wheelchairs spoke of the normalcy of emotional and physical needs. One respondent wrote: "I have been paralyzed for four years. Women ignore me completely. I am lonely and yearn for female attention." Another spoke of depression. The letters expressed pain generated by the inability to satisfy normal social and sexual desires in standard socially acceptable ways. The person from the escort service had a medical background and was prepared to deal with a handicapped client on a ventilator.

Handicapped persons suffer because even when sex is recognized as normal, foundational Abrahamic concepts provide it with a disrespectable aura greatly restricting expression. Just as with priests, the handicapped have no sanctioned opportunities to meet these needs. The harshness and lack of compassion for those unable to do so is reflective of social ambivalence toward sex and possibly the widespread nature of sexual difficulties: if you are struggling with it yourself, it can be harder to be sympathetic and compassionate with others.

The preceding conundrum points out the value of professional sex work. If sex were considered normal, proper, and acceptable, the needs of physically handicapped persons would not be seen as repugnant and routinely thwarted. Sex workers, in contrast to the moralizers who think those unable to meet these needs should just suffer, could reasonably feel they are providing a valuable service. This is a situation in which we need to recognize realities and reject religious garbage.

Priests and the handicapped are groups within the population. Teenagers are going through a life stage we all experience. While suppression of priestly sexuality harms them and those dependent on them for religious and emotional support, we all take a turn being harmed by "abstinence only" sex education. Culture in the United States, dominated by religion, presents abstinence as the only acceptable policy with reference to sex. Europe, on the other hand, teaches responsibility and birth control. As a result of these policies, the United States has a teenage pregnancy rate four to five times that of European countries.

The teenage years are not the only life stage associated with difficulty meeting sexual needs. Following the death of a spouse or a divorce, we are distressed, emotionally needy, and make even worse decisions than normal. New relationships under these circumstances are referred to as having begun "on the rebound" and are subject to elevated failure rates. It is not just a matter of us being a mess. The other person will arrive with their own problems, agendas, and baggage. Full, multifaceted relationships may be ideal under most circumstances, but there are times in life when just being taken care of would be a good thing. A sex worker can provide both companionship and sex with no demands other than money.

Consideration for these circumstances is not found in the Bible. Sirach 9:3 states, with no listed exceptions: "Do not go near a loose woman." Religious assertions are that the selling and buying of sex is always socially destructive. Prostitution was unintentionally decriminalized by Rhode Island for six years. An attorney discovered the exact nature of the law's changed wording in 2003, and the state reinstituted an effectively worded ban in 2009. The six years during which prostitution was inadvertently legalized serves as a demonstration of the effects of following religious admonishes. For starters, statewide reported rapes declined 30%, and statewide gonorrhea rates declined 40% (who would have predicted that?). Also, apparently a lot of people want to buy sex, and a lot of people want to sell sex, as these activities increased with legalization.

Both Amnesty International and the United Nations recommend legalization of sex work. What actually happened in Rhode Island with decriminalization provides support for these opinions. The 30% decline in reported rapes was not accompanied by a comparable decline in other criminal activities. As sex workers could now more readily report rape, a higher percentage of reported rapes would be expected with resultant statistics showing an increase. It is counterintuitive that it dropped. This raises the possibility that, with legalization of sex work, rape actually did decline. Once again, religious advice from an ancient civilization appears to be bad advice.

With sex work above ground, sex workers would be less subject to criminal exploitation and violence. Another aspect of legalization would be to make human trafficking more visible and more subject to control. We all remember what a big success Prohibition was. Making something illegal that a lot of people want opens opportunity for organized criminal activity. Legalizing sex work would make it safer for everyone involved, reduce criminal opportunity, and increase the difficulty in concealing human trafficking. Recall that, as previously stated, Amnesty International and the United Nations both promote the legalization of sex work.

Due to the embedded nature of religion in our society, many consider sexual activity to lack congenial aspects and always contain at least some elements of profanity. This aura of disrespectability results in the manufacture of reasons justifying a rejection of sex for money, even under circumstances in which it would be the kindest and most humane of activities. One argument against any sex for money transactions states that people only enter the field because of financial desperation. The concept being that if you had real concerns about people, you would want to keep sex work illegal.

Along the preceding lines, people only enter the field because of financial desperation, however, a lot of people would quit their jobs if it were not for financial desperation. A difference between doing sex work and working in a fast-food establishment is that with sex work you can very possibly make enough money to live on. Those making the "financial desperation" argument typically believe the problem can best be ameliorated through the jailing of participants, and not the provision of an adequate minimum wage.

With the patriarchal nature of ancient Abrahamic society, women are often the focus of sexual repression. Boys have always been forgiven for "sowing wild oats." Back in the day, the girls were to be stoned to death. Sexuality in men is seen as natural and masculine. In women, it is deviant and slutty. Women are, as a rule, evaluated differently from men under similar circumstances.

Religious documents establish women as second-class citizens under male domination (Genesis 3:16). Men have primacy and authority over women. Following the forbidden fruit incident, Eve is explicitly informed by God that Adam rules over her (Genesis 3:16). Given the symbolic nature of the Adam and Eve story, Adam's primacy serves to establish male dominance as an aspect of human nature. This view is not confined to the Adam and Eve story, but is found throughout the Bible in both Old and New Testaments. In the first letter of Paul to the Corinthians (I Corinthians 11:2-8), found in the modern "Christian Lite" New Testament, we are reminded that women were made from men and created for his sake, and that, just as Christ is the head of man, the husband (the term husband is in and of itself indicative of the man's role as overseer) is head of his wife. Thousands of years after its origins, Abrahamic religion continues to perpetuate male hegemony in contemporary societies.

The Bible presents a story depicting females as unsuitable for full societal participation. While some people consider these stories to be the definitive word of God, many others consider them to be merely harmless and cute. They are neither factual nor cute. They are socially harmful and damaging. This lies behind the necessity of woman's struggle to play a full role in society. And, as previously discussed, foundational documents may blink off but are not weeded out and are, therefore, available for resurrection by those seeking a return to what they see as the true faith.

The maintenance of a religiously justified secondary status for women is an issue of enormous contention for contemporary societies. In the United States, women got the right to vote when my grandmother was a young woman with children. There are countries today in which women are not permitted to drive cars or go anywhere without a male family escort. And, as previously discussed, women in some societies, are murdered for engaging in behavior seen as inappropriate due to religious-based gender stereotypes.

In the modern industrial or postindustrial world, women's struggle centers around equal opportunity and performance evaluation in the face of gender stereotypes. Competence should determine job opportunity and performance evaluations. Underlying gender stereotypes, however, influence both. These stereotypes may be underground, unrecognized, or just assumed to be real. For men in power the stereotypes protect their social position.

One specific concern centers around equal pay for equal work. Women quite frequently are paid less for doing work the same as or fundamentally equal to that for which men are receiving more money. This is facilitated by policies in some industries and businesses requiring that pay not be a matter for public discussion, thereby permitting inequalities to remain unobserved.

Justification for gender inequality rests on a foundation of stereotypes presenting the view that women are simply, due to their nature, less capable and competent than men. Abrahamic religions have, for thousands of years, supported these myths enabling ongoing male hegemony. The successful struggle for increasing female educational and work opportunities is shattering them. Rational, scientific, evaluation of female performance finds no support for gender stereotypes purporting female inadequacy.

Arguments have been presented supporting the notion that the Abrahamic religions do not function to produce a nicer society: as previously indicated, the areas with the highest levels of religiosity are also the areas with the highest levels of crime against both persons and property. Additionally, the Abrahamic religions promote harsh punishments for forbidden behavior.

Adam and Eve engage in a forbidden activity and they were expelled from a park designed especially for them. Of significance is that not only are they punished, but future generations are also punished for their behavior. No one will ever be permitted to return to the garden, women will always have difficult child births and be subservient to men, and men will forever need to farm for a living. It is as if someone is sent to prison for a criminal act that they commit and their grandchild will also be punished for their bad behavior.

If someone engages in blasphemy, has unauthorized sex, or generally in any way shape or form annoys God, the punishment frequently involves a violent death. Humanity has, by means of the fruit of knowledge, acquired an understanding of right and wrong. If someone does something wrong, they know it and this justifies harsh punishment.

Because, based on foundational Abrahamic documents, harsh punishment for bad behavior is appropriate and necessary, levels of religiosity should be positively correlated with high and harsh rates of punishment. And, it is. Pew Research measured religiosity by asking people if they believe in God. Gallup did so by asking respondents if they considered religion to be important in their lives, and of church members how often they attended church. As previously indicated, the Southeastern section of the United States evidenced the highest overall levels of religiosity. This region, based on the Federal Bureau of Investigation's Uniform Crime Reports, also has the highest overall rate of bad behavior with reference to both property crimes and crimes against persons.

The Southeast, furthermore, has the highest percent of its population locked up. Here we have the most religion, the worst behavior and the most people in jail or prison. It is not just within the United States that evidence for a positive correlation between religiosity, bad behavior, and punishment is found. Europe has less religion, lower crime rates, and a smaller portion of its population locked up. Germany as an example, in comparison to the United States, has a smaller percentage of its population identifying as Christian, lower crime rates overall with one sixth the homicide rate, and an incarceration rate of less than 100 persons per 100,000 compared to something in the neighborhood of 500 per 100,000 for the United States.

The Abrahamic religions inform us that we know right from wrong and that harsh punishment is appropriate for bad behavior. Unfortunately, emphasizing social control by means of punishment has its own problems as is made apparent by the high crime and recidivism rates in our lock-them-up society. It is also easy enough for the punishment itself to become bad behavior. It is the ultra-religious, such as Pasteur Swanson in this country, who want homosexuals to be executed, and the highly religious in the Middle East who are managing activities such as suicide bombing, all of which can reasonably be classified as bad behavior.

The argument has been made that religiosity is declining in the United States and that it will continue to do so. Just as overall levels of social change are not merely continuing to occur but are ramping up, it appears that the decline in religiosity is not merely continuing, but also increasing in speed. British statistics published by Britain's Office for National Statistics indicate that less than half of the population in England and Wales now consider themselves to be Christian. Christianity is now a minority religion in England and Wales.

Data compiled in the 2021 British census for England and Wales, show that over a decade, from 2011 to 2021, persons reporting themselves to be Christian fell by 17 percent and the number checking the "no religion" box increased by 57 percent (*The Economist*, December 3, 2022, page 51). Fewer people are seeking the knowledge they desire and need from the past. Religion is a voice from the past. The myth has been that God informed ancient people about truth, how their societies should be managed, and how their lives should be run. Because there is no God, people were just making this stuff up. Different societies came up with a different Word. Perhaps God changes His mind a lot or needs a good press secretary.

Science is an ongoing struggle for the acquisition of the realities we need to understand ourselves and manage our societies and our lives. Religion and science do not mix and are not compatible. They provide different answers for everything: our morality was either God given or arrived via our primate ancestors. There are either two distinct genders or something best presented on a continuum. Chapter 2 will present the scientific struggle for reality.

CHAPTER 2

Science
A Pursuit of Knowledge

After looking at what the Abrahamic religions have to offer in the way of ideas concerning human nature and how People's Land should be managed, we now look at science. The most dramatic difference between religion and science is that while religion looks to the past for knowledge alleged to have been conveyed to preceding societies by supernatural entities, science looks to the future and the pursuit of new knowledge to be added to what we think we already know. A further dramatic difference lies with the treatment of knowledge. Religion sanctifies and holds on to its ideas and concepts. Knowledge is considered to have arrived via communiques from God delivered by certified prophets, with acceptance demanded as an act of faith. Science, in contrast to religious exhortations for faith, mandates the questioning of everything. Whereas religious structure is brittle and rigid, nothing in science is held to be too sacred or too important for revision or rejection. While the contemporary Abrahamic religions have never countermanded the three-thousand-or-so-year-old claim that God created our species and did so in His image, there are no current cosmology schools adhering to an Earth-centered solar system. Religion looks to the past for faith-based truth. Science looks to the future for the unveiling of reality by means of observation, experimentation, and thought, with room for revision and innovation.

Religion presents itself as a storehouse of eternal truths, while a perusal of scientific history reveals a plethora of modified beliefs and junked concepts. Scientists construct models or paradigms of how the area of interest is structured and functions. Alterations of ideas accrue as new knowledge requires either incremental change or a model's outright rejection. With incremental change, the new data adds to what we already know, or at least think we know, while requiring only adjustments to the current model. We may already know a bit about cancer, but ongoing research should add to that knowledge. Introductory textbooks typically portray science as advancing smoothly in this manner.

However, in addition to smooth incremental change, scientific advances may occur in jolting fits and starts with what Thomas S. Kuhn referred to as revolutions. A scientific revolution occurs when knowledge is uncovered discordant with that in the discipline's contemporary model or paradigm. Because the new information does not fit, the paradigm must be discarded and another constructed to accommodate the new information.

Cosmology underwent a scientific revolution when the heliocentric model replaced the geocentric. The geocentric model held that everything orbited Earth. With the advent of the telescope, new information began to accumulate that could not be accounted for with the Earth-centric model. Objects were seen orbiting Jupiter, a full range of phases were observed for Venus, and so forth. It became increasingly apparent that something was wrong with the current model and that a new one had to be created into which the new incoming data would fit. As indicated by all the fuss and dust stirred up with the cosmological paradigmatic shift, these things do not happen smoothly or quickly.

I was fortunate enough during my lifetime to witness a scientific revolution take place in geology. When I was a kid, geology's standard model held that the continents were permanently fixed in place. This was certainly a reasonable assumption given their size. Over a period of several decades, however, accumulating evidence involving continental shapes, geological formations of one continent matching those of another, and information concerning mid-oceanic ridges could no longer be made to fit a fixed continents model. Over this period of time, a new continental drift theory was concocted and adopted in spite of the fact that it had once seemed preposterous.

One of the many interesting things about paradigms is the way they limit what we can see. Anything beyond the boundaries of the paradigm tends to be invisible. The East Coast of the Americas, fits with the West Coast of Europe and Africa. If you were to have mentioned this in a geology class in the 1950s, the comment would have been greeted with scorn. Presenting the idea that continents were floating about on a semi-molten substrate, as is held by the contemporary geological paradigm, might have qualified you for medication.

Recall that in the preceding chapter on religion, it was pointed out that ideas may blink on and off like twinkle lights in response to a collection of other interrelated concepts. The Biblical support of slavery is currently invisible due to a widespread and strongly felt rejection of the institution. In general, concepts not fitting within the interrelated details of a paradigm tend to be and remain invisible. The current view is that acceptance of something as widely condemned as slavery could not possibly be found in the Bible, even though it is readily available for any reader. A jigsaw puzzle relationship between continents could not exist and therefore was not visible. World views of both religion and science emerge from a brain with an evolved structure that functions the same regardless of whether dealing with religious or scientific concepts.

Another important point is that a trashed paradigm may or may not be something that, with thoughtful contemplation of the available information, is manifestly wrong. The Earth certainly does not feel as if it is spinning and hurtling through space. If I jump into the air, the Earth does not move beneath me as would be expected of a moving wagon should I do so while standing on it. The sun, moon, and stars can be seen to move across the sky. Given the circumstances and available knowledge, when the geocentric model was dominant, it may have been the best and most realistic assessment. On the other hand, about the ongoing paradigmatic shift taking place in sociology, the previously unquestioned idea that human behavior and society are fully and totally socially constructed may be deserving of contempt.

A further point concerning model or paradigm change is that it does not occur instantaneously like the click of a light switch – the shift in geology models took place over several decades – nor do they take place without a lot of sturm and drang. People have built careers on the old model and typically go into denial concerning the value of the new. How could the continents possibly be floating around? If you are in the middle of life and in the middle of a hard-won career thought of as quite an accomplishment, it is difficult to come to terms with the idea that what you believe to be true is verifiably not true. Max Planck recognized the difficulty experienced by those immersed in a discipline undergoing dramatically changing ideas by observing that science advances one death at a time.

The struggle to evaluate reality is both difficult, personal, and emotional. I remember an incident, early one Christmas morning when I was in the second grade and standing in our living room with my parents. One of them – I cannot remember which – pointed out our big picture window and said something to the effect that Santa's sleigh was just disappearing into the distance. I thought I saw the damn thing as it flew out of sight and said so. My parents then informed me that Santa was not real. I felt like a jerk for believing their prior statements to the effect that he was, and not just believing them but deluding myself into thinking that I had seen some sort of airborne, apparently Santa-piloted contraption disappear into the distance. I was profoundly influenced by this experience. After all, if Santa was a crock and I had bought that, what about everything else I was being told and was accepting as true? I became obsessed with trying to evaluate the ideas, opinions, numbers, advice, conjectures, reflections, judgements, impressions, beliefs, sentiments, concepts, views, and notions swirling around me like pieces of colored glass in a kaleidoscope. I began trying to use the enormous load of crap inundating me – much like pieces of a jigsaw puzzle – to construct a coherent picture of myself and the world around me.

In addition to overcoming resistance within the discipline, new ideas need to spread through the academy in general, and eventually leak out into the larger society. Just about everybody thought the Earth was in the center of the solar system and the continents were fixed in place. If it took decades for people inside and outside of geology to accept the idea that the sun was in the center of the solar system and that the continents were floating around, it just makes sense that it will take centuries for people to get used to something as personally important as that they have more in common with chimpanzees than with a god.

The religious ideal is for devotees to have faith. The scientific ideal is for practitioners to question. The reality is that both the religious struggle to have faith and the scientific struggle to question present enormous difficulties for people. We do both religion and science with the same device: the human brain. The brain is a bodily organ, just as the liver and skin are bodily organs, functioning in a manner determined by a structure that is the product of millions of years of evolution. A human brain does not perform, as much as we might like to flatter ourselves that it does, in the manner of some sort of black-box producing pure rational thought.

Our most challenging and important concern may be to understand how the brain's structure influences the way we think and how we see our world. Going back to Thomas S. Kuhn and his *The Structure of Scientific Revolutions*, he claimed that we perform scientific thinking paradigmatically. My contention is that we do all our thinking – scientific, political, and religious – paradigmatically because that is the way our evolved brain structure works.

We have collections of ideas and information concerning structures in the cosmos, the interplay of capitalism and socialism appropriate for running a country, the origin of our species, how families should be structured and run, and everything else. When new information is acquired, we attempt to fit it into our current paradigm or world view. If it does not fit so well, we attempt to cram it in anyway, or dismiss it. In cosmology, we invented epicycles to explain how what we were seeing could conform to the current geocentric paradigm. Resistance to evolution continues in some quarters to this day, and so it goes. The human brain functions paradigmatically because of its evolved physical structure. It does not function rationally.

The argument will be made, farther on in this work, that the pressure driving the evolution of the human brain forced it in the direction of enhanced community solidarity and coordinated community activity, and not some sort of general intelligence. The idea being that we dominate this planet because we compete as solidified, coordinated groups and not as individuals. Not because we are just so brilliant. We create and accept a paradigm within our group that fosters coordination and competitive advantage when interacting with neighboring groups. We accept aspects of the group world view that, if we were really wise and thinking, would be rejected. Members of political groups accept group-sponsored beliefs that appear incomprehensible to outsiders, but that because those in the group widely accept them, they foster winning elections and power. In the sciences, paradigms remain accepted longer in the face of incoming ideas until the evidence against them is simply overwhelming.

Why is it that, with religion, we flatter ourselves as having been created by, and having attributes in common with, a god; and in our scientific thinking, label ourselves *Homo sapiens*, or as being wise and sagacious. Both religion and science kowtow to our hubris and are in agreement that we are indeed very special, if for different reasons. This concurrence has been accomplished despite a history of one war after another, accompanied by a scattering of genocides. Ironically, even though we are obviously either or/both godly and brilliant, we may be doing prep work for a nuclear World War III.

Returning to the central issue of this chapter, science was applied to the concepts of human nature with the 1735 edition of *Systema Naturae* in which Carl Linnaeus, a Swedish botanist, placed people and apes in the same category. Linnaeus constructed the binomial system for classifying all living things that remains in use today and in which, 1758, he included people. He is the scientist who labeled us *Homo sapiens*. Even though his work clearly implied commonalities with other animals, its implications were not spelled out, and Linnaeus remained primarily below the public radar. It was not until a century later, with the work of Charles Darwin that science explicitly began dealing with human nature in a way that caught the public's attention.

Darwin, in the 19th Century around the time of the United States Civil War, published *On the Origin of Species* (1859), *The Decent of Man* (1871), and *The Expression of Emotions in Man and Animals* (1872). These works did not merely imply but clearly stated that people had descended from apes, shared emotions and their expression with other animals, and removed us from the realm of the gods, placing us instead into the natural world. When Darwin's work was first published just about everyone in Europe, Britain, and the United States accepted the Abrahamic canon that humans had been created by God, in His image. Humans and all other life were thought to be an "as is" Godly fabrication.

There was, nevertheless, a sense of uncertainty and concern among educated people over discordant loose ends. The remains of unknown animals were turning up in an assortment of places. This had led the scientist Georges Cuvier, in the second half of the 18th Century and before Darwin had published, to challenge the view that all animals that ever existed still did. When Thomas Jefferson launched the Lewis and Clark expedition in 1803, one of the hopes was that it would find living creatures responsible for bones discovered in Kentucky that would eventually turn out to be those of mammoths.

The puzzle pieces available to concerned and educated people in the middle of the 19th Century just did not fit nicely together. Darwin introduced the argument that contemporary animals had evolved and had not been created by God in their present form and that fossils were the remains of extinct species and not of monsters or creatures that would eventually be found alive. The controversy he ignited, especially with reference to our species, rages on today.

Ethology emerged from the European acceptance of Darwin, while the classical conditioning of Russia's Pavlov served as a foundation for behaviorism and social sciences in the United States. The difference between Europe and Russia versus the United States lies in the underlying philosophical perspectives that permitted Europe to more readily accept the biological aspects of human nature.

Russia and the United States were both birthed in revolution and steeped in ideals of human perfectibility. With its desire for a biologically unconstrained reality, Russia took to Pavlov and Lysenko. Pavlov's promise was for manipulability and perfectibility of people through conditioning by means of stimulus and response. Lysenko promised the manipulability of all biology to include such things as wheat growing in the far north.

As with Russia, the United States was imbued with the desire to have society, and life in general, unfettered by biological constraints. While we fortunately did not accept Lysenko's claims, U.S. psychologists Watson and Skinner pursued research advocating that all behavior, of both human and non-human animals was a matter of conditioning and learning with no biological input. Their ideas were absorbed by the social sciences community and have only recently been rejected, with the field of sociology even now struggling to do so.

Darwin set in motion the modern study of the role biology plays in animal (our species included) behavior and social organization known as ethology. This new discipline looks at behavior from an evolutionary perspective with reference to how it facilitates survival through adaption to the animal's environment. In addition to concern with a particular creature, ethology views similar behaviors in a range of animals. Thus, the stage was set for the study of animal behavior in a cross-specific setting. This perspective recognizes that people have an evolutionary past in which we share ancestors with other creatures. Because we are animals too, and share an evolutionary past with other animals, human behaviors, such as aggression and bonding, can reasonably be studied and compared cross-specifically.

With the development of ethology, religion and biology were offering dramatically different stories about human origins and nature. A clash between ethology and religion was certainly to have been expected, but keep in mind that we invented the god in whose image we fancied ourselves to have been created. Our species wants to be special. When science declared people to be another species of ape, it did not just collide with religion. Human hubris found it quite intolerable for us to have things in common not with God but with apes. The social sciences sprang to the rescue of human exceptionalism by mutinying and fracturing science into biological and social components.

If we were going to be related to apes, it was essential for us to remain as distinctive as could be managed. Psychology, sociology, and anthropology saw to this in at least two ways: first, the existence of a common ancestor for our species and another ape was placed so remotely in the past as to permit the argument that commonalities were nonsensical; second, the behavior of other animals was argued to be entirely a matter of biology while ours was declared to be fully socially constructed.

When I was a kid, the hominid/pongid split was placed somewhere around 14 million years ago. This was distant enough to make it a reasonable argument that our behavior had nothing to do with that of other apes. Currently, a common ancestor for humans and chimpanzees is thought to have existed about 5 or 6 million years ago. While this remains an enormous expanse of time, the results of ongoing primate research, viewed within the perspective of an evolutionary time frame, calls into question the claim that chimp and human behavior are entirely unrelated.

As far as ape behavior being exclusively a matter of biology while ours is wholly cultural, chimpanzee studies have found tool use that varies from group to group and can only be explained by ascribing it to culture and learning. Fishing for termites with twigs, for example, is commonplace in some groups, occasionally seen in others, and absent in still others. While termite twigs and other chimp tools are a long way from electron microscopes, the fact that termite fishing behavior is learned behavior varying from group to group does end the hard and fast rule that only people have culture.

While it had been conceded that nonhuman animals can have culture, the other end of the specialness argument, that biology plays no role in human behavior and societies, remained in effect. At its height, John Locke's *Tabula Rasa* theory exemplifies this perspective. He made the claim that humans were, behaviorally speaking, a blank slate at birth with no biological predispositions. All human behavior, including that associated with gender, was strictly a matter of socialization and learning.

Anthropologists, such as Margaret Mead, argued that people learn to produce and manipulate facial expressions to either inform or deceive others. The contention being that what a smile or frown means, and when they are to be employed, is fully a matter of learning and varies from society to society in contradiction to Darwin's position set forth in *The Expression of Emotions in Man and Animals*.

The psychologist Paul Ekman, in the 1960s, took issue with the social constructionist position, arguing that his research supported the notion that facial expressions were the same and their meaning recognized among all peoples everywhere, whether members of an industrial society or foragers. We might manipulate or manage them, but manipulations, to be successful, depend on the targeted party knowing what a smile means.

Ekman's work was immediately and strongly denounced by other social scientists. The fear was that if human nature existed, as implied by universally recognized facial expressions, biology played a role in human behavior contrary to the social constructionist or standard model of the social sciences.

CHAPTER 3

Religion or Science Who Knows?

Are we godly or ape-ish? The Abrahamic religions argue that we are Godly because their God created us in His image and sharing some of His attributes. Science presents a story in which we are apish, share a common ancestor with chimpanzees, and share an assortment of behavioral traits as well. If one is correct, the other is false. There is no common ground in which we could possibly be both godly and apish. To argue that we are a godly species of ape would require a dumping of the Adam and Eve story and its replacement by a tale of God starting with the creation of pond scum and working from there to us.

There are people who believe some in the "Divine Creation" story and a little bit in the evolution tale. They are cherry picking their beliefs dependent on what suits them at the moment. There is either a god hanging out somewhere and running the show, or we exist purely by means of natural selection. There is either a Heaven kind of place with God and his Buddies in residence, or all we have is People Land. One way or the other. Take your pick.

If you happen to like men in charge and prefer women as second-class citizens, do not accept birth control, think everyone who is not firmly either male or female is willfully behaving badly, and consider climate change to be an overblown issue, the God deal is for you. If, on the other hand, you think white guys do not deserve any special privileges, want all jobs to be open to all people based solely on individual abilities, that if you are married to someone and become dissatisfied it is just fine to dump them, that whoever you want to have sex with is OK as long as everyone involved is a happy camper, that children at drag shows are not a problem, and that climate change fully justifies potentially wrecking the economy: science is the deal for you.

Why is it worth anyone's time to be concerned with all of this? For one thing, giving up all our guidance from the past can be like being set adrift in the ocean with a hurricane coming. It may seem like a world in which morality and values are spinning out of control. What is next: the family disintegrating, women running everything, sex with robots, a negative income tax in which if you make less than a certain amount you are given money, clitorises the size of penises? From a conservative point of view, unhindered and unhinged progressives will turn America into some sort of Alien Hell.

For progressives, on the other hand, the beliefs passed on from an ancient civilization are often seen as out of touch with reality and discriminatory. They believe that with science and contemporary thinking, we can do a better job of recognizing how things actually are and how we should run the show.

An area of general agreement lies with science seen as more accurately ascertaining physical realities than the thinking found in ancient societies. Some of us may argue that the Earth is flat, but not that many. Disagreements typically arise concerning aspects of human nature and what these mean for society. What gender differences actually exist, for example, and is a percentage of the population just naturally homosexual?

Abrahamic societies were patriarchal and condemned women to second-class citizenship and subservience. These beliefs have been passed on to contemporary civilizations and find acceptance with a significant portion of their citizens – especially the guys. The result of these old ideas is that women today face discrimination when attempting to occupy many societal roles. As an aspect of this discrimination, women are evaluated and judged differently than are men. Whereas a man taking charge of a situation might very well be judged as appropriately responsible, a woman under identical circumstances may be judged as bossy.

Frequently, the discrimination women face is of a potentially deniable sort. Women are more frequently referred to by only their first names, whereas men more typically are referenced by their last or full names. While this may seem insignificant, it informally infantilizes female political candidates. Kamala Harris is more frequently referred to as Kamala than is Trump, for example, referred to as simply Donald.

Abrahamic religions consider their god to have created two distinct genders. There are males and females with nothing vague or in between. Anyone not conveniently ensconced within one or the other, such as homosexuals, annoys God who, as found in foundational documents, considers it appropriate for them to be stoned to death. Here again, a society's god reflects the beliefs within that society. Being stoned to death clearly constitutes discrimination.

Abrahamic gender concepts survive in contemporary societies. Today, in some societies, being gay remains punishable by death. Much of the modern industrial world has moved beyond this with homosexuals open to all career options and same-sex marriages. A significant portion of the population in even modern nations, such as the United States, however, remains hostile to homosexual people being granted standard citizenship rights. As this is being written, a homosexual man may be the most qualified to be the Democratic candidate for vice president. He may not, however, be placed in this position due to a concern as to whether or not the fact of his being gay would result in an election loss.

And then there is the matter of birth control and abortion. Today, in the United States, many conservative people support a political platform attempting to render abortion unavailable. A portion of this group would go further and also make any chemical form of birth control unavailable. Ideas from ancient civilizations continue to have an impact in our contemporary world.

To be conservative is to want to conserve what we have and possibly go back and recover some of what has been lost. To be progressive is to want to change and delete aspects of contemporary society with the intention of making improvements. Conservative philosophy looks to the past, and frequently to religion, for moral guidelines concerning right and wrong and how society should be run. Progressives generally lean heavily toward science. It is typically those of a conservative philosophical point of view who desire to keep Abrahamic religious practices in effect.

Judaism, Christianity, and Islam are the contemporary descendants of the religion found in ancient Abrahamic societies. They are of considerable import in today's world and remain relevant for many people. Conservatives typically value and adhere strongly with their concepts. The result is that conservative ideas are frequently associated with religious precepts and are lent weight because of this.

If a conservative individual and a progressive individual disagree over some point the conservative argument typically carries additional weight due to its religious associations. The progressive person may state that her or his ideals have been given extensive and serious consideration, but the conservative individual can point to their idea being in agreement with religious beliefs and therefor the "Word of God." Ideas found in the Abrahamic religions carry weight because of their religious embedment. It is not simply that Abrahamic societies believed this or that, but that they had these beliefs presented as if by their God. This was what God held to be true and proper. It is not just me. This is how God wants things done.

The concept of a living and powerful God goes beyond aspects of human nature. For some, God is viewed as running everything. From this perspective, there is no need for concern with climate change as God will take care of this, along with everything else. An individual, with whom I was talking, actually informed me that there was no need to be concerned about climate change because God controls everything and will protect us from anything bad happening.

Going back to the question of why it is important to be concerned with whether we are godly or ape-ish: how we run our societies depends on this decision. If we are godly, we will apparently consider the female half of the population to be inferior to the male half. Those not firmly ensconced in either the female or male gender categories will face discrimination. Adequate birth control may be denied, shoving women back into the 1950s. Forget about abortion.

Religion has great press. In reality it is harmful. Recognize the need for an improved world. If our efforts are not successful, at least we will have given it our best effort and not have lived constrained by what an ancient civilization believed to be true and presented as if it had originated with their invented god.

CHAPTER 4

Cooperation
An Evolutionary Theme

All living things compete for the resources necessary to establish and maintain their existence. This competition can take place either on an individual basis or as a cooperating member of a group that competes for assets. Many kinds of bacteria and some species of fish serve as examples of those that go it alone. Even the occasional mammal goes it mostly alone, except for sex. Cooperative behavior, or sociality, in which individuals join together and compete as a group, began soon after life evolved with some forms of bacteria living in colonies, and continues to this day with the social insects and our own species having achieved the greatest levels of complexity and coordination.

When Earth formed about 4.5 billion years ago, it was a hot mess. Half a billion years later, things were still pretty rough, but the place had apparently cooled down enough for the residual heat to sponsor natural chemical processes, resulting in the beginning of life. The first lifeforms were small, simple cells comparable to their modern bacterial descendants and labeled prokaryotic because they had no nucleus. It was the colonial lifestyle of these ancient bacterial progenitors that permit us to know of their existence: their colonies accumulated sediments that fossilized into what are now known as stromatolites.

Some species of prokaryotic bacteria, after living in clusters for 2 billion or so years, were cooperating on such an incredibly complex basis that they had invented a new form of life. Tiny, formerly independent prokaryotic bacteria had borged into a much larger and more complex type of cell labeled eukaryotic. These had a nucleus along with an assortment of other specialized internal structures referred to as organelles. Free-living prokaryotic cells had lost their independence, physically merged, and developed specialized functions to become elements of a new category of sociality and life. What had once been free-living bacteria had become the organelles of this new entity. Evidence of the organelles' previously free existence is provided by their having kept DNA from this prior state.

Because sociality is an evolutionary theme, certain of the new eukaryotic cells repeated the stages of advancing social complexity that their prokaryotic forebearers had followed. While some remained living in isolation, others became colonial. In about a billion and a half years, some members of the group-living eukaryotic cells had repeated the process of specializing and merging as the prokaryotes had done in the manufacture of their eukaryotic selves. This time, the result was the creation of a new, multicellular form of life. Multicellular life emerged many times, with the end result in one instance being plants and in another animals, demonstrating sociality's inevitability as an evolutionary theme. Social complexity provides competitive advantages, resulting in evolutionary transmission.

The individual organelles of eukaryotic cells, once independently survivable prokaryotic bacteria, are now specialized parts of the new eukaryotic lifeform with no possibility of sustaining life on their own. On the next multicellular level of complexity, previously independent eukaryotic cells have become specialized as liver, kidney, muscle, brain, and whatever cells with zero capacity for life on their own. At each evolutionary stage, what was once an independent organism specializes as a part of a more complex entity and loses its independence. Chuck one of your healthy liver cells out into the grass, and it is dead in a matter of seconds. If they go rogue and start reproducing on their own, it is called cancer.

Remarkably, supraorganic species echo the same diversification of function as simpler lifeforms. Just as multicellular organisms, such as humans, have specialized groups of cells called organs that do particular jobs, supraorganic ant species have members specialized for performing particular functions (workers, soldiers, and such). Just as our bodies have reproductive organs, these ant colonies have members (queens) specializing in reproduction. And, just as none of your body cells could make it on their own, neither can individual members of supraorganic ant entities.

We now have prokaryotic, eukaryotic, multicellular, and supraorganic levels of life, evidencing the reality and survival value of the evolutionary theme of sociality. With multicellular life, sociality is achieved by two means. Invertebrates, such as the aforementioned ants, do so by means of hardwiring. Their behavior is primarily controlled through genetics with a minimum of environmental input. Insect behavior, from formation of simple colonies to creation of the supraorganic, is genetically programmed to include responses to environmental input such as the presence of conspecifics, pheromones, and food.

Vertebrates, mammals for example, experience a softwiring method in which maturing genetic propensities mesh with social stimulation. Softwiring refers to the achievement of behavior through the interplay of genes and environment. The behavior is not directly programmed genetically, rather genes predispose the animal to learn certain behaviors in conjunction with environmental stimulation. With people, language is an example of how this works. We are genetically driven to acquire language when an appropriate level of neural maturation is attained and meshes with ongoing linguistic socialization.

Alteration of hardwiring requires biological evolutionary change. The softwired world, with behavior being a mix of genetics and environment, due to its social component, changes more readily and quickly in response to environmental pressures. On the downside, softwired-sponsored behavior tends to be sloppier and more loosely organized. Even if our species should attain supraorganic status, it may or may not continue to lack the level of coordination achieved in an ant colony.

The notion that advancing levels of cooperation replace competitive interactions is a misunderstanding of how sociality works. Nowak and Highfield rightly emphasize the importance of cooperation to the functioning of human communities. It should be remembered, however, that cooperation and competition are two sides of the same coin. Cooperation does not mean we avoid competition and meet our needs in a nicer fashion. Instead, cooperation is an aspect of the evolutionary theme of sociality and permits a group to compete as a united entity in pursuit of desired resources. Internal cooperation enhances the group's ability to effectively compete with neighbors.

Ants cooperate to the extent that they have achieved supraorganic status. If you happen to be another species of invertebrate, the advanced level of cooperation attained by supraorganic ants does not make them good neighbors. Our species, while not fully supraorganic, comes closer to achieving this status through softwiring than any other lifeform. We fully dominated our planet to the extent of having initiated a major worldwide extinction event – so much for the nobility of cooperation.

It is also worth noting that competition persists within the cooperating communities of social animals. Insects that have attained supraorganic status, with individual behavior controlled and coordinated like the cells in our bodies, are an exception to this rule. For the group-living, non-supraorganic species, individual members compete within their communities for dominance, sex, food, sleeping spaces, and so forth.

In the neighborhood of six million years ago, when chimpanzees and humans shared a common ancestor, some hardwired insects had already been pursuing a supraorganic lifestyle for millions of years. Our common ancestor had followed the sociality theme further than any other softwired creature, but was nowhere near supraorganic. Chimps more closely resemble this predecessor than we do. With the hominid/pongid split, it was our lineage that chased sociality with the greatest vigor, changed the most, and moved strongly in the direction of the supraorganic. Chimps are our closest relative, and we have a lot in common. Yet, there are enormous differences. A lot has happened to us in the last million years.

How have we diverged from chimpanzees and our shared ancestor? The chimpanzee cousins do OK, except when dealing with us. We subjugate the planet, destroying anything that gets in our way, and then some. The changes humans have undergone that permit this dominance involve both our level of sociality and our technological virtuosity. With reference to the social, even though not supraorganic, our species has moved strongly in that direction and with the exploding complexity and specialization within human communities continues to do so. Softwiring means a mix of genetics and environment, with the environmental portion of our mix undergoing accelerating change.

E. O. Wilson established defining characteristics for the hardwired supraorganic world. The softwired follows precisely in the hardwired footsteps. For instance, supraorganic insects have nest sites that are defended against predators and competitors. In our softwired human world, we have the same thing. The archaeologist Steven A. LeBlanc found evidence of defended living sites wherever he has dug. As human political organizations have become more complex and expanded in size from the camps of foragers to modern nations, their defended aspects have been retained.

Further, both the communities of ants and people have specialized members. With chimpanzees, everyone just does their own thing except for male/female differences. This would also have been true for our common ancestor. Even with human hunter-gatherers, separate functions are typically confined to gendered distinctions and activities in conjunction with individual proclivities. Humans move in the direction of the supraorganic with specializations emerging in association with agriculture, and then even more strongly with industrialization: our softwired species continues to move in this direction.

Both ant and human communities contain multiple generations, have members that provision them, and have members that sacrifice their lives and reproductive success for community wellbeing. Two aspects of the supraorganic that human communities do not have are member interactions as tightly coordinated as those of an ant colony and a specialized reproductive caste. The level of coordination obtainable by means of softwiring is unknown. With reference to the emergence of a human reproductive caste, perhaps advancing medical technology and increasing life expectancy will preclude just anyone from manufacturing babies and open a path for a small group of specialists. Who knows?

Inventive technology is found only with softwiring. Both people and ants have specialized individuals within their groups, but ants must grow their particularized equipment by means of slow and tedious evolutionary processes, whereas humans can create, fabricate, and alter their stuff within the same generation. Technology is not confined to us. Chimps have termite fishing twigs, but we have car keys and nuclear weapons. We did not invent the concept of the tool. Tool use is just one of those things we share with chimps and other lifeforms but which we have taken far beyond anything ever seen before on this planet.

Our sociality and technology are both changing at an accelerating pace. This world is not the world of our grandparents. Our world and that of our grandchildren will be even more profoundly different. Possibilities for the human future, neither of which is particularly attractive, include going Borg and nuclear self-annihilation.

Just looking at chimps and people, and trying to understand what has happened to us since the divergence from our common ancestor, it is our big brain that stands out. A popular explanation as to why we have achieved planetary domination is that our big brain just makes us so brilliant that nothing can stop us. The Abrahamic religions concur, postulating that we were created in the image of God and then went beyond just being in His image by adding Godly understanding and knowledge to our repertoire (through a healthy diet containing a variety of fruit). Then science came along and found us to be wise and thinking. It is nice when religion and science agree about something. Understanding how the human brain actually gets the sociality job done is another matter.

It is readily apparent that our brain must do something very important, as it is exceedingly expensive to manufacture and run. The something that it does, which is also readily apparent, is to make possible both human social complexity and technological sophistication. Setting aside technology; a clear, positive correlation exists for animals in general between social complexity and brain size. Lions, for example, hunt communally. This more complex lifestyle requires a bigger brain than that necessitated by the solitary life of leopards. Elephants and cetaceans also have complex social lives and commensurately large brains.

The demands of sociality require the memory, social skills, and judgement provided by a big brain. Chimpanzees live in complex societies and have the requisite cognitive equipment making this possible. While chimpanzee societies are complex, human societies are hugely more so, and it is our enormous brain that makes this possible. The question is: how? There are problems with concluding that our planetary dominance is purely a matter of black box, raw, rational brainpower: the "we are just so brilliant" hypothesis.

Our history is that of one war and genocide after another with quite a bit of additional violence on the domestic front. Also, with reference to this human proclivity for violence, we apparently have driven to extinction all other species of hominids. Then there is what we are doing to the environment, which is a lot like a dog peeing in its water dish; only dogs generally are not that stupid. In short, human behavior raises serious issues with the "ever so clever" hypothesis to explain our planetary domination.

Human dominance, on the other hand, does make sense if we view ourselves as a softwired version of hardwired ants. From this perspective, the role of our brain is to superglue human communities into entities capable of presenting a united front in the competition for resources, and not to make us wise and thinking. Ants achieve unity of response primarily through genetics, but softwired humans must do so through the interplay of genes and environment. While our communities do not have the unity of response of supraorganic ants, we are close enough that the only thing having a chance in competition with a human community is another human community. The big question is: what are the means by which our species, with its newly evolved big brain, achieves this level of sociality?

The answer is: through the evolution of softwired language, by means of which we have become the storytelling ape. We are disposed to manufacture and linguistically present stories espousing our group's exemplary morality and all-encompassing superiority in the face of the mendaciousness and inferiority of neighboring people and groups. All human groups, whether the communities of foragers, the nations of post-industrialists, or social and political groups within a society consider themselves to be superior to all other comparable groups.

The preceding is sponsored by a commonality of human nature. The Kung Bushmen, a foraging people living in the Botswana region of Africa, refer to themselves as "the real people." In the United States, we need agricultural workers, and yet immigrants from Mexico who would fill this need, are not welcomed by some members of our population because of a belief in our superiority. Imperial Britain's attitude toward its Indian subjects was best expressed by the London newspaper the *Daily Mail*: "This sort of creature has to be ruled, so we rule him, for his good and our own." Today, even though the British have generally abandoned pretensions of racial superiority, other versions of superiority survive.

The chatted-about view of one's group being superior fits nicely with sociality being a competitiveness enhancing evolutionary theme. Considering your group to be superior and all neighboring groups to be inferior sponsors the belief that there is nothing wrong with their exploitation, robbery, and slaughter. After all, they are at best inferior and possibly not even quite human. These ideas about yourself and others make it seem OK to sail all over the place and take anything you want. The Kung lack the weapons and combat capabilities to do the neighbors a la English. Would they if they could?

Religion coevolved with language, providing support for our sense of preeminence and rightness. It enhances our conviction of superiority by arguing that we are the preferred people of a spiritual world beyond the physical and the only group with intimate and particular knowledge of the will and workings of this supernatural world. It is a core belief of every faith that they are precisely in tune with supernatural forces. If this particularism did not exist, religion's unifying value and evolutionary ultimate purpose would not exist. Linguistically transmitted and religiously supported stories cement human groups into something going in the direction of the supraorganic.

Adding to this, political complexity involving the appearance of cities and kingships served to augment the contribution of religion to sociality. With this additional social intricacy, aristocracy, rulers, and kings were now considered to have been appointed to their positions of authority by supernatural forces, and perhaps even saw themselves as sacred. Those in charge could augment the level of community coordination by demanding obedience based on divine will. It was another step in the direction of the supraorganic.

In Bauer, we see that Sumerian religious canon considered the king to have personally descended from heaven. She further demonstrates early Egyptian pharaohs considered themselves to be serving at the will of God. While President Donald Trump did not go quite so far as to argue that He and His Court descended directly from Heaven, many of those occupying positions in that Court believed themselves to be in direct contact with and in possession of uncontestable knowledge as to Godly will.

The nuts and bolts of the language process include what is disparagingly called gossip. In actuality, gossip serves to reinforce community standards. It provides an ongoing critique of everyone's strengths and weaknesses, along with their contributions and costs to the group. Knowing that our pals will talk and judge our behavior helps keep us in line. Accompanying the gossip is frequently a discussion of goals and how they are to be achieved. Goals drive actions, and shared goals drive coordinated community action.

A suite of behaviors exists in conjunction with language. These include feelings of empathy for those in our group, but not for outsiders, a propensity to follow the activities and gaze of others, and at times for it to seem as if someone knows we are looking at them or we know they are looking at us.

This results from the human tendency to be constantly aware of the gaze and actions of others, facilitated by the highly visible white areas of our eyes. Paying attention to others enhances an understanding of what is going on in their big brains and is coupled with a desire to participate in and help with accomplishing goals. The end result is a solidly coordinated community. This is not raw intelligence. This is advanced sociality.

In addition to a belief in our preeminence, we possess a sense of rightful ownership of any desired resource – something shared with chimpanzees. Recall the conflict between Northern and Southern groups of Gombe chimps, in which the Southern group was exterminated and the Northern ended up in possession of valuable fruit trees. When it comes to raiding the neighbors, chimps get hormonally jacked up. We retain the hormones and emotions of chimps, and add in our linguistically transmitted and spiritually ameliorated stories. Every society has a story in which its members are the real people, in touch with spirituality beyond the physical world and the rightful owner of anything desired.

With the discovery of "The New World," Europeans took themselves to be racially superior, their God to be the one true God, and the exploitation of the new lands and the slaughter of their inconvenient and subhuman occupants presented a minimum of moral dilemmas. The Lewis and Clark Expedition assumed ownership of all entered lands regardless of the fact that indigenous societies were already in occupancy. The United States expanded westward through a mix of purchasing and imperialism. Ancient Israel seized land militarily that then became the Promised Land God had given them.

On the off chance you think we, as a nation, have attained enlightenment sometime between Vietnam and now, there are ranchers in the West who sincerely believe the public land on which they graze their cattle should rightfully be theirs. It just makes sense to them, the way Northern chimps felt the Southern territory should be theirs, and the way Lewis and Clark thought anything they walked on should belong to the United States.

On an individual level, a corporate chief executive officer making millions of dollars a year, while bottom level employees make something close to minimum wage, experiences emotions in common with a chimpanzee running with an armload of bananas. Emotionally, we consider ourselves to merit whatever it is that we can get our hands on.

In addition to the manufacture of stories that are objectively ridiculous, our big brain evolved to turn them into unquestionable truths. We are genetically predisposed to accept the worldview or paradigm of the groups to which we belong, often in the face of massive evidence to the contrary. "We hold these truths to be self-evident, that all men are created equal." When this statement was made, we were a slave-owning society. The United States is "the land of the free and the home of the brave." We are not rational. If we were rational, we would see the absurdities of our stories. Historic slavery and contemporary racism would make a lie of much of our self-perception.

Facts do not necessarily win arguments. If you happen to be a Democrat or a Republican, you incorporate underlying party perspectives and assumptions into your worldview without rationally challenging them. With this unquestioning philosophical base in place, the programs put forth by each leads them to look upon the other as being incomprehensibly daft. If we were actually rational, wise, and thinking, it would interfere with the solidarity and competitive abilities of our communities and groups. We are self-viewed as superior with the opposition seen as at least wrong, preferably immoral, and with their having commonalities with Tolkien's Orcs or the Evil Empire presented in Star Wars.

The extent to which members of a community accept absurd beliefs demonstrates that our brain processes information paradigmatically and not rationally. Evolutionary pressure has produced a brain that does community solidarity and not rational thought. An inclusion in the paradigm or worldview of the Christian religious community includes the belief that God impregnated a human female. A widespread belief in a god/human hybrid demonstrates that, for our species, accepting Pizza-Gate as real is not much of a putt.

The Republican paradigm includes rejection of climate change as a serious concern in the face of massive evidence to the contrary. A point may be reached beyond which a group's story begins to crumble. Just as the evidence for continental drift became undeniable, the evidence for climate change appears to be reaching that point. Donald Trump, a climate change denier, owns expensive property on a barrier island off the East Coast of Florida. Several more years of rising sea levels accompanied by storms may intrude on that version of reality.

In every society, we grow up being taught a story in which we are smarter, braver, more deserving, and all around real and better people than those of other societies (this does not just happen in Texas). These stories, in addition to enhancing group competitive abilities by welding a community together and assisting in their being wacked out killers, are absolutely essential for that society's wellbeing. When European economic and military might shattered the stories of contacted peoples, their societies were subject to discord and their citizens to suicide and alcoholism.

Perhaps our ultimate absurdity lies with the ubiquity of war. Human history is the story of one war after another. Steven A. LeBlanc, an archeologist, finds defensive perimeters and human remains that suffered violent deaths everywhere he digs. Contemporarily, we have nuclear-armed nations facing each other and comporting themselves much as chimpanzee groups. As Wilson points out, a nest and the group defense of that nest, is universal to supraorganic species. Our species, moving in the direction of the supraorganic, has defended nest sites.

Modern human war is the current form of collective nest defense inherited from our ancestors and shared with all other supraorganic or approaching supraorganic life forms. We have simply gone from the defense and raiding of small campsites to the defense and raiding of modern nations. While chimps only interact with neighboring communities, for us the entire world is our playground. World War III is an eventual certainty. "What me worry?"

Whether it is a group of chimps raiding a neighbor, or a group of Paleolithic foragers raiding another camp, or World War II, human groups can be both aggressor and defender. An important point is that no human group always plays one role or the other. The United States was defending in WW II and aggressing in Viet Nam. The best we can hope for is that our big, paradigmatically functioning brain, pre-adapts us for actual intelligence the way feathers pre-adapted some dinosaurs for flight.

As previously discussed, our brain serves not only to manufacture and linguistically present stories about our community's superiority and sponsorship by a spiritual world but also to mandate belief in these stories even though they are typically absurd beyond all reason: "all men are created equal" while we were slave owning/ "land of the free" while taking up the mantle of French colonialism in Viet Nam. Additionally, anything conflicting with the story remains invisible. The realities of slavery and the napalming of Vietnamese villages retain an invisibility permitting concepts of equality, liberty, and freedom to serve as the essence of how we see ourselves.

Also, as previously discussed, the Hebrew Bible and the Old Testament of the Christian Bible contain both a condemnation of homosexuality and rules for owners of slaves, thereby demonstrating an acceptance of slavery. Our society's worldview or paradigm is currently in turmoil with reference to homosexuality and decisively unaccepting of slavery. Under these circumstances, the Biblical rejection of homosexuality is relevant to an ongoing societal conflict. The Biblical acceptance of slavery, however, because it does not presently mesh with our societal paradigm, is invisible. For good Christians, the Biblical rejection of homosexuality represents the will of God, while the Biblical acceptance of slavery simply cannot be true and remains invisible.

Our paradigm determines what we hold to be true, and the stuff that does not mesh is invisible. When homosexuality is as unconditionally accepted as slavery is unconditionally rejected, biblical condemnations of homosexuality will become as invisible as biblical acceptance of slavery. That is the way worldviews (I use worldview and paradigm interchangeably) work because of the evolved structure of our brain and how the brain performs thinking because of that structure. The brain evolved this structure because of human sociality advancing in the direction of the supraorganic. It did not evolve this structure to do some sort of pure rational thought. The ridiculous stories mandated as factual serve to cement our communities together and thereby enhance their competitive abilities.

The way paradigmatic thinking works is discussed by Thomas S. Kuhn. Kuhn presents the argument that science does business paradigmatically. His position in that a science (whether cosmology, sociology, geology, or any other science) will have a collection of mutually supporting concepts that explain something. Anything that does not mesh with paradigmatic tenets remains invisible. As long as the geology paradigm held to fixed continents, any extent to which continents fit together was invisible.

According to Kuhn, science changes both incrementally and revolutionarily. Incremental change occurs when new information is acquired that fits into the current worldview. When information begins to pile-up that does not fit the paradigm, it remains invisible unless it becomes overwhelming.

Information indicating that the continents actually drifted around accumulated until it could no longer be ignored. A scientific revolution then occurred in which accumulating evidence led to the collapse of one paradigm and its replacement by another.

It is not only science that we do paradigmatically. We do all of our thinking – whether scientific, religious, or political – paradigmatically because that is how our brain evolved to process information. With reference to religious thinking, there is an inverse correlation between scientific education and religion. People with a minimum of science can simply ignore it. As the level of education increases, it becomes harder and harder for science to remain invisible. At some point, science comes to the fore and in our everyday thinking overwhelms religious concepts: a revolution in thinking has occurred in which we have adopted science at the expense of religion.

In addition to paradigmatic thinking, our brain does a fill-in-the-blanks thing (perhaps this is an aspect of paradigmatic thinking). We are generally incapable of recognizing a lack of knowledge. If we do not know something, we make up an answer anyway. The ancient Hebrews new nothing of evolution or primatology. That did not stop them from making up a story and accepting it as real with reference as to how our species came to exist. This cognitive process remains in place. Cosmologists always think that they now know how big the Universe is. The fact that belief as to its size increases as our instruments become more powerful just does not register. This appears to be happening again with the James Webb Space Telescope. The fact that we actually do not have a clue and are just being creative remains below the horizon. Once we believe something to be true, our brain structure and process locks it in as god or science given reality. Religion and science both contain a serving of crap, much like prisons contain a serving of the wrong people.

Our brain functions, as part of the paradigmatic cognitive process to produce certainty. In religion, our group knows what God wants; in politics, our group knows how everything from the family to the nation should be managed; and with science, we are perpetually on the edge of understanding everything. String theory explains the structure of matter, and expanding space concepts account for how the cosmos got so big so fast. Humble and confused just does not cut it when you are a species of ape approaching the supraorganic.

Getting back to our technological wherewithal, it may very well be that these abilities arrived spring-loaded as an aspect of our brain's big size due to evolved sociality. When it comes to the manufacture and usage of tools, it may actually be that we really are just so brilliant (Or maybe not – as we have created tools serving as weapons with which we may destroy ourselves).

We now turn to the consideration of gender in our human species, with the role of gender seen as adding enormous complexity to our nature. We have seen how the addition of language and religion enhanced our sociality and moved us toward the superorganic. An argument will be presented that gender changes occurring with our species have done the same thing.

Gender, of course, goes way back in our evolutionary history. In addition to sex playing an evolutionary role through the mixing of genetic material, gender categories frequently serve as the basis for specialized roles within social structure. For mammals, gender differences often involve males that are larger and more aggressive than females. (Exceptions abound, with females of spotted hyenas being larger and more aggressive than males.) With reference to apes, white-cheeked gibbons lack sexual dimorphism, with both males and females being similar in size, weight, and aggressiveness. Male gorillas, in contrast, typically weigh more than twice that of females.

Sexual dimorphism does not exist in a vacuum but rather is part of a suite of evolved physical and biosocial gender differences related to how the organism survives and makes a living. Both chimpanzees and humans are mildly sexually dimorphic, with chimps being slightly more so. This female/male difference provides a basis for female/male gender specializations relevant to survival advantage. With our species, as with tool use, it is not the existence of gender differences but their complexity and specifics that are unique to our species.

One of the nice thinks about human gender differences is that just as with religion and science agreeing on the wise nature of our specie, religion and science agree on the dominant role of males in society. In the Old Testament, Eve is manufactured utilizing one of Adam's ribs as starter material. The male is clearly the template from which the female is created. English speakers start with "man" and add a "wo" to get the altered secondary creature. In the Jewish and Christian Bibles, men have God-granted dominance over women, and in Islam it is not the guys who are supposed to remain virginal. All this stuff was, of course, made up by men.

Science, until very recently, was also almost entirely written by men. Male-dominated science marginalized women and presented men as the core of our societies and central to human evolution. Hunting, a predominately male activity, has been viewed as a key component for our becoming human and the driving force behind the evolution of language. It is an enduring male fantasy that language elevated hunting to the point of making it possible for men to feed everyone, thereby doing society's real work while women played the supporting role of staying in camp and raising the kids. Martin A. Nowak, in *Super Cooperators*, continues the perpetuation of this male myth, in 2011, speaking of the value of language to hunting.

Back in 1968, Richard B. Lee and Irven DeVore edited a book they titled *Man the Hunter*. It contains a collection of articles written about various societies that preserved hunting and gathering, or foraging, as a lifestyle. The title implied that men and male activities were the core of these communities and central to their functioning. In reality, the book's contents provided strong evidence that it was women and not men who were the communal core: women not only raised the children, but mostly fed everyone including the men. The title, *Man the Hunter*, is one of the finer examples of male chutzpa and hubris run amok in the face of massive evidence to the contrary.

With reference to the Kalahari Bushmen, one of the societies presented in *Man the Hunter*, Lee pointed out that women consistently provided two to three times as much food by weight, as men. This female-provided food was not simply a matter of salad but included high-protein items such as nuts and bird eggs. Men, to the contrary, only hunted sporadically and with widely varying rates of success. A man may have hunted for a week and then taken two, three, or more weeks off, with some simply not hunting. When not hunting, the men hung out around camp and live off the foraging efforts of the women.

When a man, or a group of men, has a successful hunt and acquire a large animal (for the 1960s Kalahari Bushmen, this could mean some type of antelope or perhaps even the occasional giraffe), it does not mean hungry children and stay-in-camp mothers will finally have something nourishing to eat. The women have already taken care of that. A successful hunt means party time and a treat. As Lee points out, "No one ever goes hungry when hunting fails." The product of a successful hunt is more like someone showing up with a case of wine, fabulous chocolates, or a 35-pound gummy bear.

Regardless of Lee and DeVore's *Man the Hunter* providing evidence clearly demonstrating the dominant role of females in providing the food as well as raising the young, hegemonic males continue to foster a vainglorious view of male social contributions and importance. Male authors continue to speak of hunting as not only being the predominant source of food for foraging communities but also the driving force for the evolution of language. Other hunting species do quite well without a lot of chatter.

For chimps and humans, statistically larger and more aggressive males exist in conjunction with the role of securing and defending territory, with hunting being a logical extension of these activities. Their bulk coupled with belligerence has permitted human males to construct and impose a male-centric interpretation of society. Men created the Abrahamic story and male god who manufactured a man first and then, because guys might enjoy some companionship, a woman starting with a spare male body part. Males are presented as the species template with females as a secondary and altered version. The Adam and Eve story is shown to be factually backwards by both genetics and fetal development.

The reality of a female template is demonstrated genetically by a chromosome, only carried by males, that serves as a switch to produce a male. If this switch is not present, the fetus will be a genetic female. Further, this fetus in the uterine environment, will only become a male if necessary hormones are present at the appropriate time of development. If a fertilized egg has not been genetically induced, and then hormonally influenced, a female fetus develops. The male is an evolved, altered version of the female to achieve specialized ends: namely to seize and hold territory.

The assertion that females occupy the central position within communities is supported by female/male differences in both linguistics and religiosity. With reference to language, on average, girls start talking a month or two earlier than boys, have larger vocabularies, more varied grammar, and in elementary school typically outperform boys on tests of spelling, capitalization, punctuation, language usage, and reading comprehension. This linguistic difference exists because language evolved focal to the female core serving to unite and solidify its members.

In addition to transmitting information, female language usage also serves as a type of grooming with a bonding function. Male communication, although strongly overlapping with that of the female, tends to serve grooming functions somewhat less while concentrating a smidgen more on the transfer of information. While female talk both transfers information and serves to bond the female core together, male talk remains more restricted to information transfer concerning activities such as fighting, hunting, and now sports. Male linguistic interactions certainly do make a contribution to the effectiveness of male-emphasized activities; however, it should be kept in mind that chimpanzees do quite well hunting monkeys and killing each other with only emotion-laden vocalizations.

Greater levels of religiosity also argue for females comprising the communal core. Gallup, in 2002, asked if religion was "very important." Sixty-eight percent of women responded affirmatively while only 48 percent of men did so. In its years of polling, Gallup reports finding women to be consistently more religious than men in both views and practices. Pew Research also finds women to be more religious that men in support of Gallup's conclusion.

Both linguistic abilities and enhanced levels of religiosity fit nicely with female centrality. Language and religion evolved as intertwined sociality enhancers serving to super-glue the society's females together, thereby contributing to community competitive advantage. Men have language and religion for the same reason they have nipples: as an altered version of the template, they get a serving of basic stuff.

Regardless of females comprising the cores, males dominate human societies. They occupy more governmental and business positions, and get paid more for the same work. Also, and in support of their being on the societal fringe, they do less housework, child care, and elder care than women. These discrepancies are justified by the widespread, socially embedded contention that men are better at the important, difficult, and skilled stuff (as defined by men) than are women. Male hegemony is in reality, founded in the statistically greater size of males, their temper tantrums, and greater levels of violence; not brains, performance, work ethic, or hormonal consistency.

As previously mentioned, when a female and a male sheriff's deputies each wrecked a patrol car under apparently identical circumstances, response for the male was "shit happens," and for the female "women can't drive and have no business in law enforcement." While cops may not generally be seen as on the cutting edge of progressive social change, surely those in the academic community would be. Unfortunately, a study coming out of Europe failed to support this contention. It was conducted by Friederike Mengel of the University of Essex in Britain, Jan Sauermann of Stockholm University in Sweden, and Ulf Zolitz of the Institute of Behavior and Inequality in Bonn, Germany. The study had a sample size of over 19,000 and examined student evaluations of female and male instructors.

For this study, students were told to rate instructors concerning various aspects of their performance on scales from 0 to 100. The resultant scores of female instructors averaged 37 points below those of males. Both female and male students gave lower ratings to female instructors, with males providing lower evaluations of the female instructors than did the female respondents. The selection of textbooks was one area of consideration. When the same text was chosen by both a female and a male instructor, the identical book was given a higher rating if it had been chosen by a male. The study demonstrated an embedded societal bias favoring men against women. This bias explains why women occupy fewer positions of power and get paid less than men. How does the "men are just better" hypothesis hold up when examined in the real world? Not well at all.

A 2016 Harvard School of Public Health study, with Tusuke Tsugawa as the lead author and utilizing 1.5 million medical records, found that if your doctor is a woman, you are less likely to die or need to return to the hospital for additional treatment. In another area, a 2001 study utilizing 10,000 randomly selected financial accounts by Brad M. Barber and Terrance Odean, of the University of California at Davis, found that women do a bit better with your money. So, you have a better chance of staying alive if your doctor is a woman and a better chance of making money if your investment manager is a woman. What is going on here?

Research conducted by Gideon Nave at the University of Pennsylvania and Amos Nadler at Western University in Ontario sheds light on this issue. Everyone makes testosterone, but males make more of it, and testosterone is associated with greater levels of aggression on a cross-specific basis. Nave and Nadler's work with human subjects demonstrated a further relationship: high levels of testosterone are also associated with impulsivity and excessive self-confidence. This fits nicely with the evolved male specialization of acquiring and defending territory.

Unrealistic levels of self-confidence, coupled with high levels of impulsivity, are a good thing for someone who is expected to fight to acquire and then defend land necessary for a community's survival – or contemporarily, participate in war and risk getting shot or your ass blown off. Women, being on average less testosterone-fueled than men, run less delusional levels of self-confidence and impulsivity. For this reason, women are less likely to make crazy, out-of-the-box decisions concerning your health care or finances. *Ergo*, you are better off with a woman doctor and a woman financial advisor. This is the exact opposite of culturally embedded beliefs.

According to male-sponsored mythology, women cannot be trusted because of their excessively emotional, childlike nature, and monthly hormonal fluctuations. In reality, it is men who are excessively emotional and prone to violence, possess greater levels of childlike impulsivity, and do not suffer monthly hormonal fluctuations but rather are consistently whacked out. Men commit over 90 percent of homicides, and are associated with the majority of vehicular deaths. The reality that women are not inferior to men, should not be career limited, and cannot rationally be paid less is emerging into the cultural mainstream. Going beyond this, it is only a matter of time until full reality leaks out and it becomes apparent that women are not simply equal to men but the less cognitively impaired species template and societal core.

Just as insulating feathers pre-adapted some dinosaurs for flight, our brain appears to have reached some sort of tipping point at which rational thought becomes at least a possibility. Being advantaged, males remain predominately committed to the current paradigm. An increasing percentage of women, however, have attained a light bulb moment recognizing that male hegemony is not thoughtfully appropriate, but rather has been installed by means of violence. A realistic view of gender differences points to women not merely achieving equality, but dominance.

Men will fight to retain their advantage, but social change is occurring that will ensure these efforts fail. Male hegemony is based on beliefs that men rightfully dominate society because of intelligence, comportment, and abilities. In reality, male dominance is based on testosterone-fueled aggression and violence. Social structure can be viewed as extending from personal relationships to nations. At all levels, the male ability to exert control through violence is in the beginnings of erosion but with a long way to go.

As one instance of this erosion, while I was a cop, laws were enacted criminalizing domestic violence. Prior to this, non-fatal and non-crippling domestic violence was considered to be a family matter. Just as children were, and frequently still are, subject to violent punishment, the childlike nature imputed to women, and their religiously sponsored subservience, meant that it was appropriate for a husband to beat his wife just as it was reasonable for him to beat the children. In today's world, at least ideally, assault is assault, and it does not matter if you are in love.

Gender roles and femineity are being redefined. When I was a kid, women went to women's jobs and dressed like women, and men went to men's jobs and dressed like men. Men wore functional pants and shoes that were suited to an active life. Women wore dresses, skirts, and shoes that physically handicapped them. They were supposed to be vulnerable and dependent on men. Strong, athletic women were masculine and unattractive. Societal ideas of what was and what was not feminine pandered to male dominance, ego, and sense of entitlement.

Physically competent and athletic women can now be accepted as feminine. This change has occurred in conjunction with redefining female-appropriate clothes. Women can wear pants. It is no longer mandated that they wear skirts, dresses, and shoes that physically disadvantage and make them vulnerable. This has taken place along with women assuming roles throughout society.

Women are, at an increasing rate, preparing for positions beyond the family, insuring that female financial independence will continue to advance. More women than men are now enrolled in undergraduate college programs. Medical schools currently have approximately equal numbers of female and male students. More women are entering traditionally male domains from which they were formerly excluded.

The increasing number of women preparing for careers is occurring in conjunction with an increasing number of men failing to do so. Males have always matured more slowly and later in life than females. With the world changing and people with penises no longer automatically in charge, there is growing male confusion concerning the nature of adulthood. As a result, an increasing number of men are extending adolescence. Some men, apparently not knowing how to be adults, are choosing not to do so at all, concentrating instead on drinking and partying with the guys, and participating in sports or obsequiously following sports teams.

Lionel Tiger, an anthropologist concerned with introducing social science to biological realities, recognized, back in the 1990s, that profound changes were taking place in female/male behavior and interactions. In his book *The Decline of Males: The First Look at an Unexpected New World for Men and*

Women, published in 1999, he argues that effective birth control is giving women complete control over reproduction and excluding men from the decision-making process. Under these circumstances, women gain full ownership of their bodies, and, in conjunction with financial independence gained by employment beyond the family, men are socially marginalized. Under these circumstances, women do not gain equality but become dominant and may clearly be viewed as occupying the societal core.

A good bet is that females, in the societies of our common ancestor with chimpanzees, formed the societal core, as is true today for chimpanzees and people. As previously stated, our species has achieved greater levels of sociality and moved further in the direction of becoming supraorganic than have chimps. This enhanced sociality was achieved by means of our big brain, language, and religion. A fourth, and equally important occurrence between our sharing a common ancestor with chimps and our becoming fully human, was the advent of males being brought into the societal core associated with their retention of immature or neotenous characteristics.

As previously stated, for the Kalahari Bushmen, when the men are not hunting, which occurs frequently, they lay about in camp and live off the foraging efforts of the women. This is a dramatic difference from chimpanzees in which males feed themselves. Men, living off the efforts of the women is also widespread across human societies that make a living by foraging and gardening. In the indigenous societies of North, Central, and South America, women typically foraged and gardened while the men hunted and often prepared land for gardening.

Men, living off the efforts of women, occurred in conjunction with them having been brought into the societal core occupied by the women. Male chimps are peripheral to the female core and provide their own food. Human males moving into the female core was made possible by means of biosocial processes, involving both males and females. Males experienced neotenous changes involving the retention of immature characteristics into adulthood. They became, in general, more childlike as adults, to the point of wanting to be mothered. Co-evolutionary changes occurred in women resulting in their being willing to mother the new neotenous version of adult males.

This mothering of adult males serves to secure stronger bonds between females and males, permitting the female core to more closely control and manage males. This adds to the community's unity and ability to deal with its environment as a single entity. The entire community was now involved in the standard ongoing intercommunity violent interaction. Rather than the females only becoming involved when their community was successfully being invaded by the neighbors, the women were now emotionally involved and supportive of the community's male violence with the neighbors. It was the invention of support for the team. It was in many ways the invention of cheerleading. We were moved in the direction of the supraorganic.

Kissing is an aspect of adult mothering as it is a simulacrum of the female feeding of children. Mothers, before Gerber began marketing little jars of baby food mush, would chew food and feed it, mouth to mouth, to infants transiting from breast milk to adult food. Adult kissing is an aspect of women mothering the new neotenous adult males.

Chimp lips are simply functional and fit together in the face. Human lips are rolled out and puffy. A transitional area of pinkish tissue, referred to as a vermillion border, exists between the mucus membrane inside of the mouth and facial skin with hair follicles. This lip area can have attention called to it through the application of lipstick or injection of substances such as Botox. Exaggeration strengthens the attraction. Chimpanzees will touch their lips to another individual, but human kissing can involve more contact and intensity as it is an adult female simulating the feeding of an adult male.

The signals emanating from breasts and lips are typically thought of as sexual, when in fact they advertise mothering. Guy chimps just want to get fucked. Guy humans want to be mothered and fucked. The evolution of permanently immature males occurred in association with females evolving both physically and psychologically to mother the altered adult males. The payoff being a tighter community with a higher level of sociality and another step closer to the supraorganic.

Lips and breasts are thought of as sexual due to males not wanting to advertise their neotenous nature and desire for mothering. With cigarette smoking we suck on a nipple like item that produces a white substance with the added benefit of a drug content. The mothering aspects of nursing on an artificial nipple have not only been concealed, but the behavior has been made macho. The Marlboro Man has explained that if you are going on a trip, such as a long cattle drive, you should make certain to bring along an adequate supply of artificial nipples to suck on. Smoking is difficult to give up both because of its mothering satisfactions and the cigarette drug content.

Several thousand years ago, after hundreds of thousands of years of foraging, farming began to spread through human societies (Chapter 6: "How Our Species Feeds Itself" will deal more extensively with this topic). The status of women changed drastically with agriculture. Males, who had been oriented toward seizing and holding territory for the entire community, now, with agriculture began owning land on an individual basis. The acquisition of individual parcels of land by males brought with it a desire for their retention. By extension, this entailed passing the land on to one's offspring necessitating knowing who one's offspring were. These circumstances lead to the extreme changes experienced by women in social structure and their relationship with men.

For foraging peoples, with nothing much to pass on and relationships being of variable duration, virginity and monogamy was of little importance. With farming and land ownership it was no longer a matter of relationships coming and going with only the emotional drama of breaking up and the thrill of new relationships forming. There was now a requirement to tightly control women and their sexuality to assure hard won land would be passed on to one's descendants. Marjorie Shostak makes informative reading with regard to relationships and gender prior to farming.

The instillation of virginity and monogamy for women, provided men with some idea concerning who their biological children were. To ensure this goal was met, the control of women began at birth under their father, with that authority then transferred to her new owner and husband at marriage. The extent of this control contemporarily varies greatly among societies, but to one degree or another, universally exists.

In some cultures, women are isolated within their families to the extent that to enter the public domain they must be accompanied by a male family member. In some places an additional requirement is that their person and sexuality be entirely concealed by the wearing of a black sack. In other societies, more liberal than the preceding, when in public women are required to cover their hair with a scarf. Failure to obey these demands may, and often does, result in arrest and beatings. Additionally, as if the preceding was not harsh enough, a brother may murder the woman with this being justified as an "honor killing" due to the embarrassment she has brought to the family.

It is easy to disparage the precedingly referenced social practices, but it needs to be recognized that the suppression of women and their sexuality exists universally, to include the Western World and the United States. This sort of thing has and continues to loosen up but with a great deal continuing to remain in place. When I was a kid, women's clothes, to a great extent, concealed what their bodies actually looked like. It might not have been a black sack but it did a comparable job. Women wore girdles and bras that altered their shape to meet societal demands for controlling the presentation of the female body and sexuality. With girdles, women's butts appeared to consist of a single cheek with zero jiggle. The bras presented breasts as solid, conical entities with, once again, zero jiggle. Women's bodies appeared to have the rigidity of department store mannequins.

Currently, it is somewhat acceptable for a woman to display a small portion of her breasts, but nipples must remain totally concealed. Failure to obey the societal nipple mandate would bring embarrassment and shame to family members, much as failure to cover the hair with a scarf would in other societies. I have been in the presence of a man who stated that if a women exhibited her nipples it would be disturbing to him. In this society, at least, it is not acceptable for a brother to kill a sister who is doing something disturbing and not representing conservative and religious tradition.

In the United States it is shameful for a woman, and distressing to her family, should she not shave her armpits and exhibit the presence of armpit hair. In addition to shaving, it is also expected that she will employ deodorant in these areas. Armpit hair distributes pheromones. Under arm deodorant eliminates the pheromones. With no pheromones or means of distribution, a communicative aspect of female sexuality has been eliminated. This is societal control of the female body and its sexuality. It is in the same category as requiring a woman to wear a totally covering black cloth sack.

Controlling female sexuality is, of course, necessitated by the need to prevent male bad behavior. Almost without exception, if men behave badly the source of this behavior lies with women: it says so in the bible (Genesis 3:12). If guys had adequate self-control, when it comes to the fathering of children, women could go around naked and use vibrators in public. Women must bear the burden of controlling their sexuality because expecting men to control theirs would place an unfair burden on them: la de da. Women covering their nipples, for example, is necessary because the penultimate event for a man is to nurse on a nipple, followed by penis-in-vagina sex. If women do not do the control job no one would have any idea of what kid was whose.

Society is no longer structured around agriculture. We have moved into an era in which a very small percentage of the population produces most of the food by industrial means. Pre-agrarian social circumstances are reemerging in conjunction with women becoming fully functional members of society.

Contemporarily, groups of women in scattered places all over the planet, are struggling to undo the control and restraints placed on them with agriculture. These controls are the foundations of patriarchy. The goal is for woman to own their own bodies and present them as they see fit. Young women in Europe and the United States typically do not wear clothing that deforms and rigidifies their bodies so as to hide their natural appearance and sexuality. Clothes are becoming more of a celebration of the body and its sexuality.

In some countries women are fighting the requirement to cover their hair with scarves. In the United States, and other societies, some women are rejecting societal requirements to modify their appearance by shaving armpit hair. A "free the nipple" movement has begun in which women affirm the right to present their bodies as they see fit and not submit to patriarchally sponsored societal demands. These and other demands on how women present themselves are in the same category of repression as requiring women to go around in a big cloth sack with only their eyes visible.

Male societal dominance is collapsing. Women are gaining financial independence and control over their reproductive processes. Marriage and being the possession of a husband is no longer essential for the existence of a family. Women are assuming more governmental positions and playing an increasing role in industry and business. Domestic violence is no longer ignored. Change is rapid and feeding on itself.

Circumstances are going beyond the elimination of those demands and limitations introduced with farming. In preagricultural times females and males were both involved in decision-making processes. Collapsing male influence may be occurring beyond the disintegration of patriarchy. Women may not simply be achieving equality. Society may be turning over with women becoming dominant. Female authoritativeness may be the next step in advancing sociality with the gender occupying the societal core assuming control. A lessening of the male role in societal management with solidified female control brings us a step closer to the supraorganic.

Males have been brought into a female dominated societal core and are being mothered in conjunction with neotenous processes. Males, however, remain a modified version of the female that evolved to acquire and defend community land. Even under contemporary circumstances, males continue to execute the endless wars that characterize the history of our species, with no end in sight. Female dominance, coupled with neotenous males, may be a step in the right direction, but it is apparently not enough to end the practice of war. Essentially, when it comes to violence between communities, which under current circumstances means violence between nations, people are the same as ever with the consequential change being the advent of nuclear weapons. Where do we go from here?

CHAPTER 5

Origins
The Ape Brain Rules

The overarching concern of this work is, given the nature of our species, how we should conduct our lives and our societies. Even though no general agreement exists concerning human nature, I argue that science is better than religion regarding the acquisition of relevant information. In Chapter 5, the concern is with the extent to which older, prehuman brain parts, influence the more newly evolved outer brain layer that we typically perceive as in charge and rational: to what extent do the older structures we share with apes, other primates, other mammals, and other previously evolved life forms, impact human behavior.

Previously the concept of cooperation, as an evolutionary theme, was introduced and the idea advanced that human societies are enormously more complicated and farther along this path than those of any other softwired species. The thinking is that the evolution of increasingly complex social behavior occurred in conjunction with the shaping of our brain, with the product of the brain's structure and functioning being of primary importance to an understanding of human nature and how our lives and People Land should be managed.

This chapter will promote the idea that our brain does not do some sort of pure, black box, rational thought. What it does do is paradigmatic thinking because of a structure evolved through natural selection. Additionally, the new, distinctly human part did not replace the older but serves rather as an addition to it, somewhat similar to adding a second story onto a one-story house. The old stuff is still chugging away underneath the new, and serving as a source for agendas and behaviors.

If the brain had not evolved but rather been manufactured by a god, its configuration and workings would probably make more sense. As when engineers design a car and are under no obligation to keep anything from prior vehicles, a god would have been under no obligation to keep any of the old ape, mammalian, and reptilian stuff, and would probably have chucked out at least some of it. Evolution does not work that way. With evolution things can be modified or shrunk, but not conveniently eliminated. If vehicles evolved rather than being engineered, a new motor could not just replace the old but would need to be bolted on top of it. Whales still have tiny, vestigial hind leg bones buried in all the fat. Humans still have an appendix.

The structure of bodily organs determine how they function; or do what they do. Just as our liver does not produce a random assortment of chemicals, our brain does not do some hodgepodge of thinking. Its structure functions to produce paradigmatic thought characterized by internally consistent, but frequently ridiculous, stories concerning our group's specialness and superiority in comparison with the neighbors. In addition to concocting these stories, this thinking promulgates their acceptance within the group as established fact.

Paradigmatic thought evolved because of competitive advantage given by promoted cooperation. The crazy stories and their acceptance serve to unite and solidify community members through a shared sense of both belonging and superiority. Additionally, a view of the neighbors is promoted as at least sub-human and with any desired resource of theirs as being rightfully ours.

The fact that the neighbors are doing exactly the same thing escapes us: we both see ourselves as the good guys with the others as orcs.

Humans do not face the world as individuals. Human groups are coordinated in their dealings with outside entities. Although not having attained the supraorganic status of an ant colony, they approach closely enough so that the only thing capable of competing with a human community is another human community. Other animals might run faster or have bigger teeth, but a Paleolithic human community was like a modern soccer team with spears: nothing fucked with them. We do not dominate this planet because we are brilliant, but rather because our groups face the world as a package due to achieved sociality through paradigmatic thought processes.

In United States politics, Republicans and Democrats have competing versions of reality that are accepted by their adherents as an unquestionable package deal or paradigm. In 2006, Al Gore pointed out the relationship between increasing levels of carbon dioxide and global warming. If our brain did some sort of pure, black-box thought, this would have met with widespread acceptance. Gore, however, was a Democrat, and the idea of global warming became associated with and a part of the Democratic worldview. Because the Democrats were the outsiders to the Republicans (much like a neighboring group of chimpanzees to our group of chimpanzees), paradigmatic brain processes required Republicans to reject the Democratic-owned concept in disregard of its obvious reality. We do not run this planet because we are wise and thinking. We run it because softwired paradigmatic brain processes produce a level of group coordination approaching the supraorganic.

Because we process politics, religion, and science with the same cognitive equipment, the answers we come up with in all three are the end product of paradigmatic thinking. We know intellectually that science entails uncertainty, but in actuality scientific concepts are accepted as established facts the same as religious or political ideas. With all three, we do not simply add in information as it becomes available. We build a story that fits the data we have at that particular point in time and hang onto it in the face of any accumulating discordant information until it becomes painfully obvious something is wrong.

When an established paradigm is abandoned due to an accumulation of discordant information, and a new paradigm is created incorporating the new information, the new, as was once true of the old, is now conceived as being precisely in touch with reality. With science, religion, and politics, we generally see ourselves as knowing just about everything or at least as being on the edge of it.

Thinking we know just about everything fits right in with labeling ourselves sapient. We delude ourselves into thinking we are wise while remaining blissfully unaware of our paradigmatic brain processes and the role played by the old, underlying ape, mammalian, and reptilian brain equipment in our agendas, desires, and behaviors. Our rationality is an illusion.

While growing up I became convinced that people were not rational or wrapped all that tight. They typically projected, or at least attempted to project, a façade of themselves and their families as being under control and moving through life in a reasonable manner. Observations would reveal an assortment of difficulties and chaos. My own family being no exception. My dad continually struggled unsuccessfully with his excessive weight. An adequate supply of food that would have been beneficial to his health was available, but he ate too much of everything. He appeared to have no thoughtful control over what or the quantity of it that he ate. His problem eating behavior appears to be widespread. Why?

Also, with reference to my dad, he was a very likable and socially skilled guy, but apparently not that smart. The "likable and socially skilled" part being evidenced by him frequently having sex with women other than my mother. The "not that smart" assessment arising from his frequently getting caught in these activities by my mother.

There is nothing rational about eating unhealthy food and overeating. There is nothing logical with having uncontrolled sexual activity. On a societal level there is no explanation as to why a sapient nature would sponsor a never-ending series of wars and genocides. An understanding of human behavior necessitates an acceptance of prehuman brain parts providing drives and agendas with the new typically making up reasons to justify what the old wants.

Additional complexity is layered on by behaviors having both immediate and ultimate causation. For an animal to keep its genes in the pool, it must mate and reproduce. Animals do not have sex because they are concerned about the future of their genetic material. If something needs to be done to assure species survival, there must be an evolved and separate ultimate and immediate causation. We, and other lifeforms on this planet, mostly do not have sex to reproduce and achieve genetic survival (ultimate causation), but because we are horny and driven (immediate causation).

Food, sex, and violence are three behavioral areas that are widely associated with problems and are strongly influenced by older brain parts. Citizens of the United States are getting fatter, sexual behavior is a major disruptive factor in relationships, and intra- and extra-group violence has remained the norm throughout human history. The roots of these problems lie in our evolutionary past. Their management requires a recognition of human nature and paradigmatic brain processes. Human problems in the areas of food, sex, and violence will now each be given consideration.

Food

Back in Paleolithic times, fat and sugar were hard to come by, with our ancestors driven to seek them out. Now, with the old drives still in place and fat and sugar readily available, we consume such excessive quantities that we are having obesity and diabetes epidemics. According to the World Health Organization (WHO), the number of obese children and adolescents increased tenfold over four decades. The National Health Study, conducted in the 1960s, found 24.3% of adults in the United States were overweight. A further National Health Study, in the 1980s (2 decades later), found the percentage of overweight citizens in the US had expanded to 33.3%.

Getting fat is contributory to contracting diabetes. The Center for Disease Control (CDC) estimates 26 million people had diabetes in the United States in 2010. Four years later, in 2014, the number had risen to 29.1 million. The American Diabetes Association states that diabetes kills more United States citizens than AIDS and breast cancer combined. Both being fat and having diabetes are associated with heart disease, and heart disease is the leading cause of death for both women and men in the United States.

According to *Cooking Light* magazine, dinner plate size has increased 22% over the last century. The *Food & Brand Laboratory* at Cornell University concurs with the claim concerning a dramatic increase in plate size, further arguing that this is a critical factor in widespread weight gain. Apparently, we just shovel in whatever is in front of us. If the plate is bigger and holds more food, we just gobble it down with no thoughtful consideration given to nutritional needs. Dinner table behavior is not characterized by rational self-control.

Research by Lauri Nummenmaa of the University of Turku in Finland finds that eating results in the release of opioid brain chemicals. These "feel good" chemicals reward us for eating with their release not related to enjoyment of the food. He found that chowing down on some sort of unappetizing "goo" produced the same level of release as good-tasting pizza. This serves as further evidence for an evolved drive for food consumption. If a behavior is necessary for survival, it will be driven subconsciously. We remain driven and chemically rewarded for feeding behavior when we no longer need to work so hard at it.

Problems with our eating habits are not just associated with drives but how our brain is structured and functions. Our paradigmatic thinking means that we create and believe stories about our food and eating that are as out of touch with reality as those concerning the exceptional nature of our groups. There are United States citizens who argue that high calorie and nutritionally poor fast foods, with resultant weight gain, are not a problem. According to contributors of *The Fat Studies Reader* the real problem is not obesity, but a negative attitude toward it. The diabetes and heart disease associated with being overweight provide a counterpoint to this argument.

As a further example of the effects of paradigmatic brain processes, an industry once argued that tobacco and smoking were not bad for your health, and currently the sugar industry contends its product is not a health problem either. Gary Taubes, in his book *The Case Against Sugar* presents a different point of view. He argues that, just as with tobacco, sugar is profoundly harmful. Given the paradigmatic nature of our brain, sugar industry representatives do not need to see themselves as evil and trying to get rich selling a harmful product. Our brain permits convenient delusional thinking in the face of massive evidence to the contrary for both sugar industry personnel and consumers.

Harm done by the way we eat is obvious. Unfortunately, the way we eat is driven by agendas of older brain parts that in the past promoted survival. We delude ourselves into thinking we are rational. We are not rational, and our weight control efforts typically fail because of agendas promulgated by brain tissue we share with apes. Perhaps, as science and technology permitted us to get into this mess, by making fat and sugar so readily available, they can help us deal with it.

Chemical compounds can exist in two forms that are the mirror image of each other. Sugar comes in what is known as a right-handed and a left-handed version. The right-handed version is the only one found in the natural world. Left-handed sugar must be manufactured. As a result of its artificiality, human digestive enzymes do not interact with it. It passes untouched through the body. This is not to advocate lives totally devoid of self-control, but perhaps we can manufacture versions of fat and sugar that will permit us the luxury of indulging some of our primitive urges without getting fat, diabetic, and dying from it.

Abdicating personal responsibility by looking to science for an easy fix will almost certainly leave many readers feeling uncomfortable. We did, however, acquire the problem as a result of technological sophistication. Perhaps it is not unreasonable for science and industry to play a role in providing at least a partial solution. We are eating ourselves into obesity, diabetes, and heart attacks. We need all the help we can get.

Sex

While we may self-servingly or self-destructively deny unpleasant dietary realities, the situation with regard to our sexual behavior is even worse. Whenever a moment of clear thinking can be managed, it becomes obvious we should not eat ourselves into obesity, but nothing is apparent when it comes to sex. It is not even unquestionably accepted that a sex drive is part of human nature, or even that there is such a thing as human nature.

The sociological standard model denies all biological influence on human behavior. According to Emile Durkheim, a founder of sociology, human behavior results from social circumstances and is socially constructed in its entirety. There is no human nature and there are no biological drives. If this were just a case of an academic discipline being out of touch with the real world, it would be one thing, but a failure to recognize the existence of human nature, with a sex drive component, is widespread enough to do considerable damage. As previously discussed the Catholic Church is struggling with the results of this failure.

Catholic priests have no acceptable sexual outlets and, just like the mole, their sexuality pops up somewhere or other. Allowing priests to marry would help, but in and of itself would not be sufficient to deal with the problem. In addition to permitting marriage, both priests and their parishioners would benefit from premarital sex, birth control, and masturbation being dealt with realistically. As priests are human beings, bad behavior would still occur, but the extent of it would be more in line with that of the larger population.

As previously discussed, priests are not the only group facing an inability to meet normal social and sexual desires in standard socially acceptable ways. In addition to sex workers, there are sex surrogates, trained in the psychology and physiology of sex, and capable of helping people with physical, emotional, and mental health issues, engage in sexual activities and deal with sexual difficulties. Other physical and emotional issues may respectably result in seeking of professional help. The same should be true for sexual situations.

Not that long ago we lived in a world of fanciful notions about human sexuality. Then along came Alfred Kinsey, as previously discussed, a biologist and specifically an entomologist with a specialization in gall wasps. His personal experiences, growing up in the early years of the 20th. Century, called his attention to harm perpetuated by unrealistic ideas concerning human sexuality. Accepting ideas about what we were supposed to be sexually, in the face of what we as individuals knew we actually were, frequently led to the conclusion that there was something wrong with us.

Kinsey studied human sexual behavior with the same rigor he had applied to wasps, finding enormous disparity between what we thought was true and what was actually true. As an example of the disparity between accepted wisdom and the data Kinsey and his associates gathered, masturbation was widely believed to be abnormal and subject its practitioners to debilitating and even fatal illness. Kinsey's data, contrarily, reported a masturbation rate for males of 92% and for females 62%. Very few of these people appeared to die or go blind.

Another area in which Kinsey's data conflicted sharply with popular opinion concerned homosexuality. While same-gender sex was widely believed to be exceedingly rare, Kinsey's data found 2 or 3 percent of the population to be exclusively homosexual. The even bigger surprise was the number of people who were bisexual or at least had one or more homosexual experiences. For the male interviewees, 37% reported having had a homosexual experience resulting in orgasm. Results from the female sample indicate 13% of those respondents had experienced one or more orgasms from same sex contact.

When it comes to masturbation, pre-marital, post-marital, extra-marital, homosexual, heterosexual, sex with other species, sex with dolls, the beginnings of sex with robots, and who knows in the future maybe sex with space aliens, people behave more interestingly than ordinarily given credit, and they spilled their guts to Kinsey and his associates. In addition to many people finding the results distressing, a concern for their accuracy existed because interview subjects were volunteers and not chosen randomly from a target population as required by sound scientific procedure: perhaps only the weird and perverted had consented to be subjects.

Relative to this concern is research that has blossomed since Kinsey's pioneering efforts. The Institute for Sex Research was founded in 1947 at Indiana University, where Kinsey and his associates were working, and continues its mission to this day of casting light upon human sexuality. Also, sex research is now widely pursued at universities and medical schools. In addition to greatly expanding our knowledge, results from this ongoing work have shown Kinsey and his associates to have been in the ballpark with the majority of their conclusions.

Ongoing efforts support Kinsey's work concerning the commonality of masturbation. An article in the *Archives of Pediatrics & Adolescent Medicine* states, "in studies of older adolescents and adults, masturbation is nearly universal among males and reported by a majority of females." Research conducted by University of Virginia Medical School faculty members Cynthia Janus and Samuel Janis confirms the complexity of homosexuality explicated by Kinsey. They estimate about 9% of men and 5% of women have "occasional homosexual relationships." While Kinsey's data present higher percentages, they include those with a single experience, whereas the Janus' dealt with ongoing "occasional" behavior. Kinsey's sampling methods are of realistic concern, but ongoing research provides substantial support for his results.

We have come a long way in knowing how people actually behave. The big question now is: what do we do with this knowledge? Distress has frequently led to its rejection as not factual or it simply being ignored. Additionally, it may be accepted as fact but condemned for moral or religious reasons.

Boy Scout leadership knows jerking off is normal and will not make you go blind, but they still do not like it. The Boy Scouts of America's Executive Committee issued a statement in 2013 stating, "any sexual conduct" "by youth of Scouting age is contrary to the virtues of Scouting." Jerking off is a sexual activity; it is not virtuous, and the Executive Committee decidedly does not approve.

In association with the preceding, the sexual abuse of an assortment of Scouts by a medley of Scout Masters has been documented. Dr. Janet Warren, Professor of Psychiatry and Neurobehavioral Science at the University of

Virginia Medical School, in a five-year contract with the Boy Scouts of America, researched Scout files and requested volunteers to come forward concerning sexual abuse experienced in scouting. She collected data concerning abuse occurring between 1944 and 2016 and found over 12,000 reasonably substantiated examples.

Even though it is recognized that good Christians might masturbate, religious arguments exist indicating they should struggle against the practice. The Christian Apologetics & Research Ministry has this to say about the matter: "the desire to be pure and holy should move the Christian to avoid it," CatholicBridge.com reports, "chastity means no sex with self or others before marriage" and calls masturbation "intrinsically and gravely disordered."

With reference to actual harm, Sandra Glahn and William Cutrer, in their book *Sexual Intimacy in Marriage*, claim masturbation damages marriage, arguing that if a young woman masturbates, it interferes with later marital sex. Kinsey's data found the opposite to be true. In response to claims such as the above, he reported "pre-marital experience in masturbation may actually contribute to the female's capacity to respond in her coital relations in marriage." Females who had experienced orgasm prior to marriage were more readily orgasmic in marriage. This information is not something the religious set wants to hear, so they reject it.

Even though orgasm through prior masturbatory experience facilitates orgasm in marriage, complexity is generated by masturbation more readily generating orgasms than penis-in-vagina sex. Glahn and Cutrer, once again in their book *Sexual Intimacy in Marriage*, make specific reference to vibrators, asserting, "I heard from women who started using vibrators but then found it a challenge later to receive pleasure from the 'real thing'."

All women require adjustment to the "real thing" whether or not they have masturbated with or without a vibrator. The masturbatory and vibrator set is, in reality, further down the road to happy sex, and also better informed as to what gets them off with the greatest efficiency. Kinsey's female subjects reported achieving a higher rate of orgasms through masturbation than they did through sex with a male. Glahn's argument boils down to advocating ignorance. If you have never masturbated or used a vibrator you do not know how effective these procedures and this equipment can be in achieving orgasm.

Recognition that "do-it-yourself" methods and electrically driven paraphernalia can generate timely results, entail none of the potential emotional stress of activities involving another person or persons, and necessitates minimal clean-ups is in no way intended to negate how fabulous a penis can be for everyone involved. It should just be recognized that if you are late for work, going through a bad breakup, absolutely could not endure another load of laundry, or just feel like it, jerking off or applying an electrical device to your crotch may be precisely the thing to do.

CatholicBridge.com – credit where credit is due – provides a religious perspective of benefit in evaluating the gravitas of various ways of achieving happy sex. They argue that birth control, by eliminating the reproductive value of sex, makes couples sex the equivalent of masturbation. This may point us in the right direction. If we are having sex for pleasure, and not aiming for reproductive success, perhaps our sexual interactions should be more in the nature of masturbation. Penis-in-vagina sex could be stripped of its holy-grail status but retained as part of a repertoire, should participants so desire and have both a penis and a vagina available. The idea is to do what you like, as long as you like, and eventually get on with getting off.

Valuing the penis to the point of permitting its ownership to interfere with the best possible sexual experiences is not unheard of and goes back in history. Freud, in the company of others concerned with mental and emotional functioning, believed that a mature female would experience vaginal orgasms.

With a guy's dick being the very center of meaning and pleasure for him, it just makes sense that it would be what is happening when it comes to women as well. The reality is that the clitoris, in its entirety, is what is happening when it comes to female orgasm, and the nerve endings do not somehow migrate to the vagina as a female matures as would be required by psychoanalytic theory.

The fetus, in the uterine environment, grows either a clitoris, a penis, or something in between, from the same lump of tissue depending on hormonal presence and sensitivity to it: they are equivalent organs. A penis gets direct stimulation from moving in a vagina. A clitoris consists of more tissue than the surface protrusion in the vulva. Clitoral tissue extends to and may surround the vagina and may possibly be associated with the Grafenberg or G-spot. As such, interior clitoral tissue may receive stimulation from penis-in-vagina sex. The suggestion is, however, that direct and special attention to the external clitoral tissue might be a good thing and provide the direct stimulation a penis receives during vaginal insertion.

Typically, females have clitorises and males have penises, with this equipment playing a role in how gender is constructed and sex conducted. A big problem is that conservative and religious people want everyone to fit into the two neat categories of male and female: they consider this to be God-given and not up for grabs. Reality requires gender and sexual preference to be viewed on a continuum with people all over the place and not necessarily affixed to one spot. The attempt to force conservative beliefs concerning human gender and sexuality on society is another harmful aspect of religion arising because religion is out of touch with human biology.

An underlying question is how much of gender is nurture and how much nature? What aspects of human biology sponsor gender differences and to what extent do they do so? To what extent can gender differences be reduced or completely eliminated?

The basis for human behavior lies with our brain. The differences between a human female brain and a human male brain are not obvious. An appropriately trained person can make a very accurate guess as to whether a brain had been extracted from a female body or a male body. The differences that exist overlap with commonalities of both brains. There is no distinctly female or distinctly male brain.

With reference specifically to gender differences and behavior, a high percent of females wants to have sex with males rather than with other females, and a high percent of males want to have sex with females rather than other males. This does not seem to be solely a matter of cultural imposition, but to involve a biological component. Homosexual attraction appears to be comparable and similarly to involve a biological component. In the absence of cultural sanctions, the percentages would change, but to one extent or another continue to exist because of a biological aspect of gender. Both heterosexuality and homosexuality are influenced by cultural factors but are not simply a matter of choice.

We evolved as a species with biology playing a role in gender differences. An evolutionary perspective assists in understanding influences from our pre-human past and in grasping the differences that have evolved between chimpanzees and us. The present-day communities of chimpanzees are more similar to those of our common ancestor than are human communities. The distance between human behavior and communities and those of our common ancestor and contemporary chimpanzees is reflective of our having moved further in the direction of the supraorganic through the evolution of language and religion and the greater integration of males into the communal core.

As previously argued, with both chimps and people, males are specialized community members geared toward the acquisition and possession of territory necessary for group survival and prosperity. Human males are more integrated into the female communal core than are chimpanzee males. This has been accomplished through neoteny with human males retaining juvenile characteristics and never getting over wanting to be mothered. The females, for their part, have developed a willingness to mother adult males.

A significant chimp/human difference lies with the visibility of female sexual receptivity. A female chimpanzee's receptiveness to sexual activity is clearly indicated by a swollen and reddened genital region. Human females do not exhibit such a readily perceived signal. Male chimpanzees know who is hot and who is not. Male humans do not generally have a clue. This is in conjunction with human males having been introduced into the female societal core, with the general availability of sexual activity, and the non-announcement of female sexual receptivity being a component of this inclusiveness. Human males are constantly interested in females and not just during their ovulation with relationships extending beyond and between these periods.

Money, children, and a lot of things complicate relationships, but our sexuality has got to be somewhere near the top of the list. As pointed out, lifetime monogamy and the imposition of female virginity is an invention of, and appears historically in conjunction with, agriculture. Having females marry with the beginning of reproductive possibilities and sexual receptivity made virginity prior to marriage a manageable possibility, at least for females. Males married later in life with no serious demands for virginity. Prior to agriculture, virginity was not an issue as there was minimal personal property to pass on following death and the big stuff, such as land, belonged to everyone.

Today, females only very rarely get married at puberty. Now, both females and males, with the attainment of puberty or not long thereafter, may begin having dating relationships. We then have a series of relationships, one or more of which we may formalize by marriage. What is important is that rather than one relationship beginning in virginity and lasting from puberty until someone's death, we now typically have a series of relationships that last for various lengths of time. Even when a marriage endures throughout the adult lifetime it has often been preceded by dating relationships and no longer may or may not begin with virginal participants. Additionally, it is not rare for one or more of the participants to have one or more incidents involving an outside person.

The decreasing commonality of lifetime marriage leads to the increasing angst and suffering of relationships unraveling and ending. The accompaniment of serial relationships is separation and divorce. As our understanding of human sexuality increases, no answer is found as to how to conduct our lives so as to avoid suffering. It seems we have evolved to survive and not to be happy. Learning what we are learning may not permit us to live stress free blissful lives, but it may provide us with realistic expectations to pass on to children. Teaching children to expect a lifetime monogamous relationship may be harmful in two ways. Firstly, if, as frequently happens, a marriage of theirs falls apart, they may have a sense of failure when in reality they are simply experiencing a widely common situation. Secondly, should their parents divorce, they may consider them to be failures, when in fact they are opening up the possibility of a run at a more successful relationship.

A problem area within marriages concerns the definition and particulars of monogamy. Looking at soft porn or flirting may or may not be considered a breach of monogamy by participating individuals. As with just about everything, gray areas exist: when does a kiss go from a friendly hello to a sexual exchange. This is the sort of thing that can only be decided on an individual basis. For the purposes of this work, a relationship will be considered to lack full monogamy status if it is the primary current relationship and one or more of its members engages in a physical, sexual activity with a person outside of the relationship.

The General Social Survey, in 2016, reported 13% of women and 20% of men as having had sex with someone outside of their marriage. Statistics concerning non-monogamous behavior, or infidelity, are fraught with potential inaccuracy. Researchers are asking about participation in an activity that a majority of people consider immoral. The subjects of these surveys are being asked about an accumulation of incidents over the entire course of their marriage: there is no statute of limitations: some may conveniently misplace the memory of a distant event. Because the marriage continues, future sexual incidents remain a possibility. To explore available infidelity statistics, refer to the General Social Survey, The Institute for Family studies, YouGov surveys, Gallup Polls, and the Pew Research Center.

While disagreement exists concerning infidelity rates, it is widely recognized that a higher percentage of both females and males are participating in these activities. University of Washington researchers, using data collected in 1991 and 2006, found infidelity rates rising across all age groups. They report women's rates in their 35-and-under age group to have risen from 12% to 15% and men's rates in this category to have risen from 15% to 20%. In their over-60 category, women's rates have risen from 5% to 15% and men's rates from 20% to 28%. Explanations as to why these increases are occurring point to multiple, interrelated causes involving both social and technological change.

With reference to social change, dating relationships are extending over a longer expanse of the life course in association with both females and males marrying later in life. Census data indicate that from 2005 to 2015 the average age at which women married went from 25.5 to almost 28 years of age. Average age of marriage for men went from 27 to almost 30 years. Additionally, marriage rates peaked at the end of WWII and have steadily declined. Barely over half (51%) of those over the age of 18 are married. Cohabiting is replacing marriage for an increasing share of the population, with some cohabitors simply putting off marriage while others say cohabiting is an end game. With reference to infidelity, cohabitors have higher rates than those who are married.

Both divorce for marriage and breakup for cohabiting puts people back in the dating pool. It is not uncommon for someone to have multiple marriages and/or cohabiting experiences interspersed with periods of dating. Dating, of course, has higher rates of infidelity than either cohabiting or marriage. Individuals have an increasing number of people with whom they have had sex. The end result of all this is a population used to having sex with an assortment of people. It is less of a big deal for them to mess around.

Contemporary medical knowledge, drugs, equipment, procedures, and prosthetic devices keep us functional farther along in life. Artificial joints, such as those for the hip, alleviate pain and facilitate physical activity. Refinement of cardiac procedures and the development of implantable devices such as pacemakers, defibrillators, and ventricular assist devices keep us alive longer and ameliorate overall quality of life. The drug Viagra, and its generic equivalents, sponsor erections and erectile firmness for older males. All of these modern medical achievements extend sexual functioning for both fidelity and out of marriage sex.

In addition to physical capability, communicative opportunity is also in ascendancy. Cell phones and computers permit individuals to talk and plan effectively and privately. With the image of the other person on a device's screen, it is, to an extent, as if that person were actually present. Additionally, internet sites exist that specifically facilitate extra-relationship sex. Technological communicative advances do not just facilitate adult sexual behavior but permit young people, via computer image, to be with their girlfriend or boyfriend in the privacy of their room beyond parental control.

Another product of advancing technology is the increasing availability of pornography or, if you prefer, erotic material. This may influence how often and in what ways we think about sex and possibly what activities we consider reasonable and acceptable. With reference to young people, pornography may be a prominent source of information concerning sex. It is a safe bet that most porn does not promote relationship fidelity.

As previously noted, religion is declining and secularism rising. Pew Research Center reports that in 2007, 16% of the United States population considered their religion to be "nothing in particular." By 2014, that rate had risen to 23%. While the "nothings" may or may not condemn infidelity, it is a good guess that the religious crowd exhibits a more consistently hostile perspective. Under these circumstances declining religiosity could serve either to reduce infidelity by improving sexual satisfaction or to take the brakes off the decline in partner faithfulness.

Serial relationships and enhanced opportunity undoubtably play a role in the increasing rates of infidelity, but by far the most profound societal changes center on gender. Behaviorally, female and male genders are converging. Females are advancing in terms of both education and employment. The percentage of females in all phases of education, from undergraduate to professional, is growing and predictive of growing participation throughout the workforce. While pursuing education and advancing in employment is external to home and family, gender convergence is also occurring on the home front. Money means power, and more financially secure and strong women happen to be, the more power they wield in the family.

Female income is rising, with about one third of women earning more than their spouses. Women with careers and money are not male dependent. Financially independent women have higher infidelity rates than financially dependent women. In addition to financial independence, women meet more people and have greater opportunities for extramarital sex when in the workforce.

Available data typically indicate that men commit more adultery than women. The General Social Survey, in 2014, found an exception to the "males cheat more than females" rule. In their 18 to 29 aged married category the females reported an infidelity rate of 11% while the males reported 10%. For this population the rates for females and males, for all practical purposes, had converged to the point of being identical.

The converged 18 to 29-year group is obviously young and early in their marriages. We do not know if this convergence will hold throughout their relationships or if they will return to the more familiar "guys cheat more" pattern. Factors argue social change is taking place and the convergence may hold. The General Social Survey reports female infidelity rate, for the total married population between 1990 and 2010, increased 40%, bringing the female rate to 7 for every 10 men.

One thought concerning rising female rates of infidelity and convergence with male rates gives consideration to the possibility that women's sexual behavior may not be changing so much but that they may be becoming more honest about it. Traditionally, both women and men valued female circumspection when it came to sexual behavior. Whereas males might brag about sexual exploits, females were behooved to do the opposite, perhaps to the point of lying. If the guys are claiming a certain amount of sexual activity, beyond that with the Palmer twins, it only makes sense that women, even though they may deny it, are involved.

Increasing female acceptance and honesty concerning their sexuality may portend behavioral changes not ending in precise equality with that of males: flip a coin and it probably will not land on its edge. Female sexuality may become more prominent than that of males. In addition to doing better as doctors and investors, once freed of repression they may also fuck more.

As Esther Perel, in her book *Mating in Captivity: Unlocking Erotic Intelligence*, points out, extramarital sex can be problematic, but it is not necessarily a dealbreaker. As a Relationship Councilor, Perel is concerned with maintaining the quality of relationships when and while they exist. In her referenced work, she recognizes the difficulties of maintaining the worth of a relationship, including what she refers to as "hot sex," over an extended period.

In addition to length of time for a relationship, raising children presents extraordinary challenges. In chimpanzee society, the males may be violent and annoying but primarily only want sex with estrus females. Human males want recurrent sex, along with perpetual mothering. The addition of a genuine infant into a human relationship injects a great deal of stress into what was already problematic. Regarding the social management of People Land, and taking into consideration all of the prior referenced complexities, it is clear that the sex drive must be acknowledged, but beyond that little is certain. Monogamy is held as an ideal by a species of ape.

The substantive changes that have occurred in the way women and men view gender and gender relationships not only impacts marriage rates but the totality of female/male interactions. It is not only marriage rates that are declining. Romantic relationships, in general, are doing so. The General Social Survey (GSS) reports that fewer people even have steady romantic relationships, let alone marriages. GSS data collected in 2004 found 33% of 18 to 34-year-olds were not in a steady romantic relationship. Data collected in 2018 – only 14 years later – show an increase to 51%. With a decreasing percentage of steady romantic relationships, marriage rates will almost certainly continue to decline.

Young people are not just having fewer relationships, they are having less female/male sex. The Center for Disease Control's Youth Risk Behavior Survey found, from 1991 to 2017, the percentage of high school students who had experienced penis-in-vagina sex had dropped from 54 to 40 percent. Rates for oral sex remained steady so it was not a change in fashion for getting off. This has been given a positive spin and interpreted as high schoolers behaving better.

Unfortunately, seen within the context of decreasing romantic relationships and declining marriage rates it may be subject to another interpretation that is not all that positive. The December 2018 issue of *The Atlantic*, in an article by Kate Julian, contains the following statement: "Signs are gathering that the delay in teen sex may have been the first indication of a broader withdrawal from physical intimacy that extends well into adulthood." It is not just penis-in-vagina sex being put off until after high school, and marriage being put off until later in life, the entire gender relations ball of wax is experiencing some level of disintegration. Young females and males are, across the board, building increasing distance.

Finland's *Finsex* survey also found declining rates of penis-in-vagina sex. The Finnish research further indicated that declining couple's sex was occurring in conjunction with increasing rates of single person sex. The data indicate that from 1992 to 2014, men who reported having masturbated in a given week more than doubled. Whack-A-Mole Theory continues to hold. We are not having fewer orgasms. We are having more of them in isolation.

Increasing male isolation is occurring in conjunction with gains in female empowerment. We have always had isolated males with physical handicaps and mental illness. As males lose control in relationships and must adapt to greater relationship equality, or going beyond that to female hegemony perhaps in association with higher incomes, a growing number just cannot or will not hack it. With sex, there are times, even when we are in a happy and fulfilling relationship, when it is just nice to do it yourself. An increasing percent of the male population is moving in the direction of a higher percentage of isolated sex, and even in the direction of only having isolated sex.

Consideration has been given to times in life when doing it yourself, or paying for sex, may be a reasonable choice. With the decoupling of genders, we are going in the direction of an increasing percentage of people finding couples sex not to be worth the effort. A biologically based drive exists sponsoring coupling as a type of bonding behavior. When males elect not to pursue coupling, it is an emotion-laden decision. Heterosexual males rejecting intimate female companionship and involved in what is known as the manosphere frequently exhibit and promote hostility toward women. The decision not to pair up produces emotional distress due to the biologically based drive for companionship with whatever it is that you happen to be attracted to sexually.

One group promoting gender separation and hostility is "Men Going Their Own Way" (MGTOW). A common driving force behind this group and others in the manosphere, is resentment toward the advances being made by women. They recognize the melting away of male hegemony and are not at all happy about it.

Couple's relationships are being affected by an increasing availability and acceptance of sex toys and equipment. Vibrators, sex lubes, male masturbators, and such, can now be found on display in some pharmacies along with the vitamins and cough medicines. The availability of these can serve to enhance a couple's sexual activities and play a positive role in the quality of their lives. Also, people who are isolated, temporarily or permanently for whatever reason, may find solace with them.

On a less cheery note, their isolated use, by individuals within a couple's relationship, may serve to increase the distance between the partners. The toys facilitate just masturbating rather than making the effort required to maintain a relationship. With their availability, they further facilitate complete, no partner, isolation. Those who would prefer a couple's relationship may find they serve as an easier alternative than dealing with another person. Sex toys may facilitate the non-partnership and isolation of individuals. As discussed, someone preferring human companionship but not successfully achieving this desired state, may blame others and become hostile to those seen as the problem.

Beyond vibrators and masturbators, sex dolls are also becoming more widely available. They have been around for a long time associated with men experiencing periods of isolation such as sailors and other travelers and known variously as Dutch wives, azumagata (Japanese for woman substitute), lady travelers, and so forth. With increasing rates of non-coupling, rather than the temporary isolation of sailors and other frequent travelers, they can facilitate permanent isolation. When it comes to couple's relationships in People Land social change is out of control.

With advancing technology sex dolls are in the beginnings of evolving into companion robots. Realbotix, a California company, wants to make something with which companionship and bonding are possible: something with which to chat while watching TV. The goal is to move the dolls beyond masturbatory equipment by means of chatbot technology, internet connectivity, and an adjustable personality. Where can this go? Put the Wright flyer next to an F35 to get some idea of the technological change that can occur in one hundred years.

Every person comes with problems, agendas, and baggage. Following two or three human relationships with bad breakups, a growing assortment of people may conceivably opt for a break, on a temporary or permanent basis, with human partnering. If something with no emotional baggage, and concern only for you, talks to you, and fucks you, you might bond with it and get weird about it: If people can love cats they can love machines.

Artificial friendships have commonalities with alcohol. Both could be a positive and enjoyable addition to life or both could wreck it. An artificial companion as an addition to a circle of friends, or a temporary couple's arrangement while getting through one of life's hard spots, could be beneficial. For the long haul, however, it may or may not work out so well. People Land is evolving into something we have never seen and are incapable of imagining.

Violence

Violence is a problem in all human societies and will continue to be so. It is primarily, but not exclusively, a male activity. It is males who commit most of the murders and other violent crimes within societies, and males who fight the wars occurring between societies. The origins of these behaviors lie far back in our evolutionary past. It would be wonderful if the people in our past were peaceful and nice, but they were not.

Human prehistory evidences violence between individuals and groups. With written history we are presented with unending wars typically coupled with mass killings of civilians. Ancient writings, as found in the Hebrew Bible/Old Testament of the Christian Bible, present their God as demanding the slaughter of defeated enemies. More recently, World War II saw the bombing of Dresden, a city lacking military significance, and the nuclear bombing of Hiroshima, another city of no significant military importance. Viet Nam witnessed 500-pound explosive bombs and napalm (jellied gasoline) dropped on villages of men, women, and children. Following Viet Nam, widespread war continues. Today it is ongoing in Ukraine.

People did not invent violence. Richard Wrangham and Dale Peterson argue that violence is not only characteristic of our prehistoric past but our prehuman past as well. Our violence goes back to and beyond the common ancestor we share with chimpanzees. As previously discussed, with our species of ape, males are an altered version of the female designed through natural selection to seize and hold territory. Heightened levels of male violence lie in our prehuman past and have an evolved biological basis.

As a softwired species, socialization plays a major role in our behavior. This makes it possible to reduce levels of male violence through management of the socialization process. Violence can be encouraged or discouraged. The extent of acceptable levels of inter-personal violence has and is continuing to be reduced. Domestic violence has been criminalized. Financially independent women will put up with less of it.

Full commitment to the reduction of violence is lacking. Controlling children through corporal punishment teaches them to solve problems through violence. The United States has not signed, and refuses to sign, the United Nations Protocol on Children's Rights, passed in 1989, because multiple states in the South continue to permit corporal punishment in schools.

Controversy exists concerning the circumstances under which reciprocal violence may reasonably be considered self-defense. When is a person required to leave the scene of potential violence, and when is it acceptable to stand your ground? What sort of threats justify what sort of responses in a person's home or business? In Texas, under certain circumstances, you can shoot someone and it is OK. Under identical circumstances in Massachusetts, you would go to prison. This reflects differences in culturally acceptable levels of violence across regions and states. Historic flow is moving in the direction of Massachusetts.

While domestically we are sweeping less violence under the rug, recognizing its existence, and reducing it, wars continue apace. War, with the advent of nuclear weapons, threatens the continuing existence of our societies and species. While societal male violence is in reduction, there appears to be no reduction in the frequency and destructiveness of war. Men initiate and fight wars. Managing the socialization input to the construction of male behavior does not appear to adequately remove their commitment to war. Apparently, biological aspects behind the male seizing and holding of territory may be so considerable as to render the manipulation of socialization processes inadequate to end male commitment to war. To end war, and avert nuclear-self-annellation, it may be necessary to terminate the altered male version of the female. It may be necessary to deconstruct masculinity to assure human survival.

Summary

We either share the wisdom of God, acquired illegally in the Garden of Eden (religious perspective), or have evolved as "wise and thinking" (scientific pronouncement). My argument is that both are in error. We are neither godly nor wise and thinking. The wise and thinking perspective holds that our newly evolved brain parts are running our lives and societies. In reality, the new has not taken over and reduced the old to an anachronism. The old continues to chug away underneath the new, sponsoring the agendas and behaviors it always has. The new, rather than superseding the old, works to carry out its directives at a level of sociality beyond that of our chimpanzee cousins.

Chimpanzees are straightforward, emotional, and honest concerning the pursuit of self-serving agendas. Chimps just go for it in an openly selfish way. Our species, on the other hand, has advanced farther down the road toward the supraorganic through the evolution of paradigmatic justifications for behavior that amounts to self-delusional bullshit. Chimps just do what they can get away with. We manipulate others to achieve our ends.

The ape brain rules. In People Land, we need to acknowledge and manage the old underlying desires for food, sex, and violence. Maybe it is this old stuff that gives life its meaning. If we actually trashed all the old and ran only with what our new brain parts could come up with, our lives might seem meaningless, with suicide more rampant.

Even though we are not rational, wise, or logical – in a moment of unjustified optimism – I suggest that we may be preadapted for rational thought the way feathers preadapted some dinosaurs for flight. Who knows?

CHAPTER 6

Food
How Our Species Feeds Itself

Life exists within a thin layer above, on, and in the surface of the Earth. We evolved in this biosphere and are a part of it. How we see ourselves, in relation to the rest of life, impacts our behavior and how we feed ourselves. How we feed ourselves impacts back on our view of our place in the natural world. A dramatic difference exists between the worldview of foragers/horticulturalists and that of agriculturalists.

Our species has been around for a couple of hundred thousand years, and we have foraged for our food for almost the entirety of that time. This way of life was inherited from our prehuman ancestors. We lived ensconced in the natural world and, although our presence impacted our surroundings, we fed ourselves by selecting from items available in that environment.

The environmental impact of foragers occurs when fruits, nuts, and seeds are dropped, possibly resulting in plants growing in new places. Seeds that can survive a trip through a digestive tract are positioned in a new location along with some fertilizer. Our species may have noticed the sprouting of seeds and deliberately spread them around: who knows?

More extensive manipulation of the environment became possible with Homo erectus, our immediate ancestor, becoming acquainted with fire around 300,000 years ago. With the burning of an area, new and more tender growth would emerge. In addition to possibly being human food, the new sprouts would attract animals that could also be eaten. Stolzenburg suggests with reference to the American Midwest: "The endless grassland of legend is likely an ecological artifact sculpted by aboriginal peoples with a penchant for fire."

Regardless of the environmental manipulation that occurred through the spreading of seeds and the use of fire, hunter-gatherers lived by foraging within their environment. This lifestyle necessitated considerable walking to search out available comestibles. It was the women who did this and provided most of the food. Distances covered and the necessity of childcare limited the number of children a woman could manage and served to limit population growth.

While most of the women were out foraging, some of the men would be out hunting. Anthropological data, however, indicates many of the men would be hanging out in camp and being useless. Back in the day, men did not help with childcare. Men have only contemporarily begun to contribute. Male uselessness would have necessitated women taking their children with them. In this manner, men would have played a role in the population control needed to maintain the foraging way of life.

It was not until twelve thousand or so years ago that we took up farming, on a here-and-there basis, and began a more intensive manipulation of the environment. With agriculture, plants are domesticated by selecting the variants perceived as being most suitable for human needs. These are then bred, producing a domesticated plant that is recognizably different from its wild ancestor.

Cereals found outside of areas in which they appeared naturally are one bit of evidence supporting the existence of agriculture. Whereas foragers lived in and harvested from a complex environment, farming works for the control and simplification of the food source. Rather than selecting items from a diverse array, farming provides monocultural areas for harvest.

Agriculture appears to have developed independently in at least six places. It was once thought that it had begun in just one place and then spread. Distances between locations of origin and the variety of crops grown at these locations argue for multiple beginnings. Multiple beginnings imply some common circumstances occurring and driving each of the origins. Rapid population growth is consistently found with agricultural beginnings. We do not understand this relationship. It may be that a dramatically growing population serves to push for agriculture, it may be that the increase in available food with agriculture fosters population growth, or it may be that a third factor is involved that drives either agriculture and/or population expansion.

Perhaps some foraging areas were so rich that minimal walking was necessary for the harvesting of food. Under these circumstances, women could have given birth to more children, thereby leading to expanding populations. This population growth would have eventually resulted in a need for more food and driven the agricultural revolution. In this scenario, population growth drives farming.

Farming produces more food in a smaller area. Some communities may have started to place the seeds of plants they deemed to have more desirable characteristics in areas where they might grow, thereby laying the groundwork for more food and the beginnings of farming. With the availability of more food and, thereby, less hiking, the brakes would have been off population control. In this scenario, farming drives population growth.

A third possibility is that farming and population growth occurred together in association with a third factor. Slingerland suggests the domestication of wheat was driven by the invention of beer. Agriculture came to exist so that more wheat would be available for beer production. Some of the increased supply of wheat could have been eaten, reducing the amount of necessary foraging. In this scenario, farming and population expansion occur simultaneously with an increasing demand for and supply of beer. Alcohol derived from other vegetative material could have facilitated agricultural origination in non-wheat areas.

Another factor playing a role in the beginnings of agriculture has to do with the local availability of plants and animals propitious for domestication. Areas with wild grains held greater potential than many other areas: already existing wheat seed heads were a start in the right direction. Early domesticates also included the other grains of rye, barley, rice, and maize. Additional domesticates included peas, beans, squashes, and grapes. Many trees produce edibles, with olives and pistachios eventually being domesticated, however, the domestication of trees was not as early on or as common as grains and vegetative plants, as they lack the selection immediacy of herbaceous annuals.

Agriculture also includes the domestication of animals. The beginnings could have involved capturing young animals or acquiring them through the successful hunting of the mother. They could have been kept as pets and/or eaten, possibly when they became larger. At some point, particularly desirable animals would have been kept and bred. Here again, what you had to start with would have played a role. Domestication occurred primarily with animals found in Europe and the Mid-East, such as goats, sheep, cattle, and pigs. Few major meat-producing animals were domesticated in the Americas due to a shortage of potential domesticable possibilities.

The first items to be domesticated were already in the diet. Archaeological research shows we were gathering wheat seeds before we began to domesticate them: we were already making bread and beer. Some societies just continued to harvest what was available, while others began to selectively maintain and plant variants. Why some did and some did not is open to speculation.

Whether a community existed by foraging or farming can be archeologically determined because of recognizable differences between domesticated varieties and their wild ancestors. Wild and domesticated cereal grains occurred together for an extended transition time. We continued to harvest the wild while putting effort into the taming process: there was no light-switch moment. This is known because of the distinctive characteristics of the wild and domesticated remains found at settlement sites. Ancestors of domesticated wheat have seedheads in which individual seeds mature at different times, and the seeds fall from the stalk when ripe.

Harvesting wild wheat is made laborious because as you are cutting a stalk, some grains will dislodge and go all over the place, and some will be ripe, and some will not. Selecting and planting seeds from the rare genetic variants that held their seed more firmly and with more regularity in ripening eventually produced domesticated wheat. The domesticated plant holds its seeds while being harvested and then requires threshing to separate the seeds from the stem. The result is a much more complete acquisition of the seeds. Farming takes us in the direction of control.

The change from foraging to farming occurred in conjunction with a dramatic change in worldview. The ancient Hebrews were an agricultural people, and the Abrahamic religion and the god they invented present their perspective, in the Hebrew and Christian Bibles, of how people fit into the natural world. The picture presented is not one of a society immersed in and extracting a living from nature. It presents a society wanting to own and control its world. In Genesis 1:26, their invented god gave them "dominion" over the total biosphere. It was all theirs, and they were in control and could do whatever they wanted with it. We were no longer a part of the biosphere. Now we were owners.

This anciently contrived version of a god and supernatural presentation of reality has been passed down, in its totality, to a big chunk of today's world.

The Abrahamic view sponsors and suits modern agriculturalists and many contemporary societies in general. Today's industrial farming pushes manipulation and control of the environment, very much in extension of Abrahamic concepts.

In addition to the manipulation of their environment through the use of fire, many foragers further alter their surroundings through gardening or horticulture. With this, the environment is altered through the insertion of a selection of plants deemed to be of value. We are then foraging from a manipulated landscape that includes the addition of selected items. It may be highly altered, but it remains complex, and we remain a part of it. Indigenous populations in the Americas typically included gardening in their foraging way of making a living.

Even though the modern world is thoroughly dependent on agriculture for its food supply, where opportunities exist, many people also forage, fish, hunt, and garden. As these activities are not essential, this speaks of ancient and genetically embedded desires surviving from the millions of years we and our ancestor species roamed the wild to feed ourselves.

It might seem likely that with the advent of farming, we would have had plenty to eat, and everyone would have been better off. Looking at skeletal remains proves the opposite. With farming, human health deteriorated. People became shorter, with their remains exhibiting increased signs of nutritional deficiencies, disease, and infections. Teeth enamel became thinner, along with more cavities and a greater loss of teeth.

Hunter-gatherers ate an enormous variety of plants and animals (I am using the past tense because this way of life no longer exists in its pure form). This diversity resulted in their nutritional needs being met. Farmers eat a more limited variety of foods with a higher concentration of carbohydrates. Foragers simply had a much healthier diet.

Farming communities are dependent on one or a very few crops. When the rain did not come, or too much came, or anything occurred that disrupted the growth of crops, people went hungry. With foraging, environmental conditions may have interfered with the availability of what was most commonly or preferably eaten, but with their knowledge of the environment and its diversity, something was always available. Farming invented famine. Recall the Irish famine occurring in association with the potato crop failure.

Hunter-gatherers lived in small groups and typically had an annual travel cycle based on what foods were available in what areas. Farmers lived permanently in one place, with larger numbers of more densely placed people. This closeness facilitated the transfer of diseases between people and the occurrence of epidemics and their associated deaths.

It was not just that people were packed more closely together. With farming, people lived in closer association with the animals they ate than did hunters. This closer association facilitated the adaptation of the animal's disease-causing bacteria and viruses to a new human host. With animal husbandry, human diseases spiked as demonstrated within the fossil record.

The movement of foragers meant they were not always drinking from the same water source. Agriculturalists, on the other hand, stayed permanently in place, often resulting in contaminated water serving to spread disease. Dehydration resulting from diarrhea caused by contaminated drinking water was and remains a primary health threat contributing to infant deaths.

Foraging generally produces an adequate supply of food, with occasional excess resulting from the successful hunt of a large animal. Growing one crop in a given area permits a big harvest when conditions are good, supporting population growth. With farming, good years and bad are intermixed. Whereas hunter-gatherers experience regularly available food, farming populations go from feast to famine.

As if all this were not enough, it seems that the softer agricultural diet does not provide the chewing resistance necessary for the development of our teeth and jaws. Ungar argues that our teeth and jaws evolved with a diet more typical of that of foragers, with food that was denser and harder and contained less carbohydrates and sugar. In support of Ungar, fossil evidence presents forager jaws that are more substantial and in which room exists for wisdom teeth. Their fossil teeth also exhibit considerably less decay.

The larger populations supported by farming expanded into forager territory and pushed them onto land not suited for farming. With any hostile interaction, the greater numbers of agriculturalists overwhelmed forager populations. The annual cycles of foragers, in which they followed food availability from one area to another, facilitated land loss by providing temporarily unoccupied territory for farmers to move into. Having, at some time in the past, been driven from a rich and diverse ecosystem, theKung now lives in the Kalahari Desert. Even under these circumstances, their diet remains nutritionally adequate with no concern for a lack of something to eat.

Humans, engaging in agriculture, have spread to occupy the majority of the planet, carrying the Abrahamic worldview with them. In Genesis 3:17, our species is granted "dominion over the fish of the sea, and over the birds of the air, and over the cattle, and over all of the wild animals of the earth, and over everything that creeps upon the earth." Unconditional population growth lies enmeshed within and sponsored by this perspective. As God's chosen people we are to multiply and our population is to expand without bounds. Birth control, or sex for fun, seriously annoys God. No other creature or its habitat is of any consequence. As Elizabeth Kolbert points out in *The Sixth Extinction*, the Earth has gone through five major extinction events with a sixth cranking up courtesy of us.

We plan on agriculture feeding us forever. In reality, we are approaching a precipice. Before modern farming, the prairies and marshlands of the

Midwest had thick soil held in place by vegetation. Bison and an array of animals lived in association with the plants. Native Americans managed this ecosystem to suit their needs: their fire kept trees from taking over. It was living with and managing a diverse ecosystem.

In contrast to the preceding Midwestern circumstances, the installation of agriculture has resulted in the loss of at least half of the prairie soil: the soil is eroding much faster than natural processes can replace it. Corn and soybeans are grown year after year on the same land. Between the harvesting of one crop and the planting of the next year, the soil is most frequently left bare and exposed to wind and rain, resulting in the aforementioned erosion.

Soil is a complex medium that sponsors plant growth. It contains both mineral and organic matter. The mineral composition depends on local geology. The organic matter consists of both dead material and living fungi and plants. With industrial farming, a mix of chemicals, which theoretically provide everything needed for crop growth, is added to what remains of the soil. Trace elements that impact the nutritional value of the crops may, in reality, be missing. In addition to providing chemical fertilizer to sponsor crop growth, herbicides and pesticides may be applied.

With erosion of the soil, the added chemicals go with it, polluting downstream water and land. After this abatement, rock particles of various sizes are primarily what remains. We are moving in the direction of growing crops in chemical-laden grit, very much the equivalent of hydroponics. With hydroponics, however, the involved chemicals are managed and do not end up being washed downstream.

Modern agriculture is driven by money and government subsidies. The chemical industry benefits financially from the sale of agricultural chemicals. Some of the profits are employed to elicit governmental policies favorable to the industry through the employment of lobbyists. One successful result is governmental farm subsidies that encourage the continuation of industrial agricultural practices. These practices lead to such things as algal blooms and dead zones in the Gulf of Mexico, with damaged fishing and shrimping industries. It also appears quite likely that agricultural chemicals directly impact the human population.

Before harvesting, animals are fed hormones and antibiotics to facilitate weight gain. The hormones include both testosterone surrogates to promote muscle (meat) growth, and estrogens to stop the animals' reproductive cycles and thereby save those resources for muscle growth. While some of these chemicals exit the animals in urine and feces, some stay in the animals' tissues and are consumed along with their meat. The accumulation of animal-derived hormones in consumers may play a role in increasing rates of weight gain and diabetes. The antibiotics, without a doubt, play a role in the development of drug-resistant bacteria.

Back in the 1980's, the U.S. government gave approval for the use of hormonal steroids, such as trenbolone, to facilitate weight gain through increased muscle mass in cattle. The thinking disregarded the possibility of their presence in animal tissues being harmful to consumers, and further entertained the notion that these large molecules, when excreted, would rapidly degrade and present little danger of accumulation in water and soil. Research coming out of the University of Iowa, the University of Florida, and the Environmental Protection Agency among others, demonstrates that these chemicals and their metabolites are not degrading as completely and as quickly as had been supposed.

Run-off from cattle facilities appears to be serving as an endocrine disruptor in wild fish populations, with, for example, females producing fewer eggs. These substances not only impact cows and fish. People shoot up with this stuff for the same reason it is given to beef cattle. Additionally, aging men take testosterone (another steroidal hormone) in hopes of remaining young and virile. Undesirable side effects are common in both cases, leading to such things as moodiness and lowered sperm count. If these chemicals are accumulating in our water and soil, it is of very reasonable concern. Under those circumstances, we are all getting dosed up whether we want to or not.

The point of antibiotics is to kill single-cell organisms. A scattering of which, both inside and outside the cow, chicken, person, or whatever, will be resistant, survive, and pass this resistance on to the next bacterial generation. As a result of this process having been ongoing since the introduction of penicillin, the World Health Organization (WHO) reports a decreasing effectiveness of antibiotics on a worldwide basis. Conditions such as bloodstream and urinary tract infections, which in the past were routinely treated with antibiotics, are now more commonly not responding due to drug-resistant bacteria. In hospitals, infected wounds are, with increasing frequency, not responding to common antibiotics.

Modern agriculture, with its chemicals, appears to be harming our general population. From its very beginnings, agriculture has sponsored the destruction of soil. Historically, farmland degrades over time. With farming, the previously existing plants are removed and the crop is planted. The removed plants had roots that held soil in place and permitted their healthy complexity. With their removal and the planting of crops, the soil erodes. The hills of Greece are so bare and scenic because ancient farming led to their prior soil cover being washed into the valleys. Soil in the Midwestern United States is ending up in the Gulf of Mexico thanks to contemporary industrial farming: big changes and the same result.

Our Abrahamic heritage leaves us with the sense that we not only own but can also manage all natural processes so as to benefit ourselves. California goes from droughts to floods. With droughts, enough water does not exist to meet agricultural needs and those of a growing population. The response is to meet perceived water needs by pumping from aquifers. Efforts are not made to manage water by such means as choosing crops that require less water or rethinking golf courses and suburban lawns.

As groundwater levels lower with pumping from deeper and deeper wells, a point is reached at which the water being extracted has increasing levels of salt and other substances such as arsenic. Eventually, the extracted water becomes undrinkable and unsuitable for watering crops. With water pumped out, the land subsides. This lowering alters surface elevations, damaging irrigation systems, roads, and buildings. With subsidence, the compacted ground can no longer absorb water to refill the aquifers should wet periods last long enough to do so.

None of the preceding is given consideration as to applicable remedial actions until a crisis level is achieved. Apparently, the growth of California's farms and cities will just keep rolling along until crops are dying and golf courses are losing their greens due to water shortages: we are, at present, attaining this point. As situations become critical, we start to think about it with offered solutions consisting primarily of arguments as to why a particular group should get whatever remains of the resource.

Historically, California occasionally experiences episodes of rain heavy enough to flood its farms and cities. In the droughts, the floods are forgotten, and in the floods, the droughts are forgotten. With heavy precipitation, the effort and money put into cropland (carrots, broccoli, cauliflower, lettuce, strawberries, tomatoes, and so forth) and tree farms (almonds, pistachios, citrus) is lost, and Lake Tulare reappears.

The sense of ownership and control of the natural world is not limited to agriculture but extends throughout the societal worldview. Spring flooding of the Mississippi River was inconvenient for human habitation and businesses. The United States Corps of Engineers would take care of that! Enormous and costly efforts have been put into confining the Mississippi within desired banks and levees. Flood control efforts are now quite good and actually work until a major hurricane comes along and overwhelms them.

The Mississippi carries a large load of eroded material from the lands where it originates and moves through. In the past, the river would enter its delta over a large shifting area, slow down, and deposit its load widely. With its canalization, it enters over a smaller area, retaining enough momentum to carry most of its load beyond the delta and into the depths of the Gulf of Mexico. Without this annual replenishment, the Mississippi Delta is shrinking. In the last 50 years, Louisiana has shrunk by some two thousand square miles. The lost land had people living on it and was interspersed by waters important for seafood production and harvest.

The lost land (this loss is ongoing) also served to distance New Orleans from the Gulf of Mexico and the land entry point for hurricanes; hurricanes lose force as they travel over land. Also of interest is the fact that most of New Orleans lies at or below sea level. In some places, it is actually 15 feet below. Further, New Orleans is subsiding while the sea level is rising. In places, this subsidence is close to half a foot a decade. This is hurricane country. The idea that we can control the environment enough to keep New Orleans a viable city is ludicrous.

The agricultural paradigm sponsors the concept that people own and control the natural world and are to be fruitful and multiply without limits. The human population continues to expand under this perspective. Our population has encroached on and damaged, to one extent or another, all ecosystems on this planet – recall Kolbert's "Sixth Extinction." How should we manage People Land so that it is both OK for us and does not wreck the rest of the biosphere?

When I was young, zoos kept animals in cages. Under these circumstances, the creatures were unable to act out genetically programmed behaviors and exhibited tendencies for self-destructive conduct such as chewing their feet or damaging their teeth by chewing cage wire. Today, zoos attempt to provide accommodations as similar as possible to those the animal would encounter in the wild. This change is in recognition of behavior being an important part of an animal, with a natural setting permitting inherently natural development. Just as genetics determines body structure, they also predispose an animal to behaviors. An animal that can act out its biological programming is less prone to self-destructive acts and easier in general, for zoos to manage and keep healthy.

The preceding also applies to our species. Unfortunately, we know less about people than we do about many other animals: major contributing factors being that we have an unrealistic perception of ourselves as rational, and are masters of the façade. We present our families and other groups with their conflicts and chaos as well-hidden as possible. Just like Monkey Island in the Philadelphia Zoo, People Land will always and forever exhibit considerable disorder regardless of how well things are structured and managed.

Zoos frequently have areas dedicated to baboons. Because zoo people are in charge, these areas are generally well managed. If the baboons ran Baboon Land – if you will indulge a silly fantasy of mine – things might not go so well. The baboon population might expand until Baboon Land takes in the entire zoo. Other animals would be shoved aside and relegated to inadequate areas. Many would die, and baboons would end up comprising most of the zoo biomass. We could hardly expect baboons to have behavior more praiseworthy than our own. After all, People Land has expanded to encompass most of the habitable areas of this planet and is currently sponsoring a major extinction event.

Our enormous population feeds itself using an agriculture that has become industrialized. In addition to replacing living soil with chemical-laden grit, animals are treated as if they were cogs in a machine. No allowance is made for biological behavioral predispositions or an environment permitting adequate space in which the animals can move around. Egg-laying chickens exist in areas suitable for a mechanical device but not a living thing. Sows feeding young pigs are confined to rectangular areas that fit them with minimal possibility for movement. Programmed behavioral predispositions are totally ignored. It is reasonable to assume that animals treated as if they were machine parts would suffer from it. If we are going to eat something, it should have a decent life.

With reference to farming and plants, crops are annual. Every year, the soil is plowed, crops are planted, and the results are harvested. Every year, a new, typically monocultural, system of plants is established. Widespread agreement exists that this is the only way our species can feed itself. Bellwood states: "Hunting and collecting entirely from the wild could not possibly support even a tiny fraction of the world's current population." He then points out that: "The development of the agricultural systems that provide virtually all of the world's food has occurred over many millennia, and still proceeds apace."

Bellwood is, of course, correct in his statement that we could not feed humanity by pre-agrarian foraging. But neither could we through early agricultural methods. The agricultural revolution pointed us down a path that now supports us. The Abrahamic philosophical system is a package deal. If we had not proceeded down the agricultural path for millennia, our population would not be so enormous. It is also of considerable significance that, rather than being forever, the road down which we travel is leading to a cliff: soil is washing away, and the biosphere which we are a part of and dependent on is being altered.

It is encouraging that in another 80 or so years, the human population may peak and begin to decline. This occurs in association with a philosophical shift demonstrating abandonment of the idea that a god made everything for us and that we should multiply as efficiently as possible. Hopefully, we have crossed a threshold and are now beginning to recognize that rather than owning the biosphere, we are an integrated part of it. It may be time to rethink how we feed ourselves.

We may not go back 12,000 years and start over, but we can begin to move in the direction of long-term sustainable complexity as opposed to annual monoculture. The suggestion is to establish integrated systems of plants and animals from which food can be harvested over an extended period of time. This integrated permaculture (IP) method avoids the energy expenditure and soil damage and loss associated with annual plowing and planting. Recall that, as Bellwood pointed out, the development of contemporary agriculture "occurred over many millennia." The development of integrated permaculture is also a long-term deal.

On the Midwestern prairies, as they were found by Europeans, grasses were perennial with deep roots that held the soil in place. With the prairie wetlands included, an enormous assortment of plants, mammals, birds, reptiles, amphibians, and insects were a part of this ongoing system. Native Americans fed themselves in this land by managing and harvesting food in a manner that was permanently sustainable (permanently sustainable by human timeframe standards, not, of course, permanently within geological time).

Our current Midwestern population could not be fed from this area even if we were capable of returning the land to the conditions found by the first Europeans. The size of our population is a central issue and must be dealt with for anything to work. And, it is neither possible nor desirable to attempt to restore the ecological circumstances Native Americans were living in. We need to assemble a malleable and research-oriented ecology in which our species can feed itself now and into the future. Forget about preserving some sort of natural environment. That is a long-lost cause.

We have impacted and shaped the environment since our very beginnings, generally in a random and unthinking manner. As our population grew and spread, we carried plants and animals with us. We brought along mice, rats, and cats. Cats may have been for the control of the mice and rats, but they also killed birds, lizards, other reptiles, and small mammals, driving many to extinction. From our beginnings to the present day, we continue to distribute large numbers of foreign species all over the planet in complete disregard of existing ecosystems.

There is now an established population of Burmese pythons in Florida. The adults of these snakes average out to around 16 feet, with some up to 18 feet and over 200 pounds. They eat anything they can swallow. Examinations of stomach contents of euthanized specimens included rodents, raccoons, squirrels, egrets, herons, ibis, coots, alligators, and other items too digested to label.

A diverse assortment of other snakes (to include other species of python) and lizards have gained Florida residency. Nile monitor lizards, reaching an adult size of around 8 feet, eat other lizards, frogs, fish, ducks, snakes, and so forth, and have also established Florida breeding populations. Tropical areas all over the planet are subject to a variety of invasive species: Guam, for example, has the mildly venomous brown tree snake that has devastated local populations of birds, mammals, and reptiles.

Many of the newly arrived residents of Florida were released by hobbyists when they no longer wanted or could manage them. Additionally, Hurricane Andrew caused the release of a wide range of captive animals from private collections, zoos, and research facilities. These included, reptiles, monkeys, birds, fish, and amphibians. Brown tree snakes arrived in Guam as stowaways on military aircraft.

Birders in Florida see parrots from all over the world. Parrots are interesting because some of them have not confined their residency to Florida, but are adapting to colder climates and are thriving in rural areas and cities around the world. As with many of the reptiles, breeding populations of parrots began with the release or escape of pets. In general, parrots are social and beautiful, with some being harmful agricultural pests. Some are inconvenient for other reasons, such as outcompeting native species for nest sites and food. Monk parakeets build nests that may be 10 feet in diameter, house dozens of birds, and short out power lines as far north as New York City and Long Island.

It is not just reptiles and birds. There are 4 species of Asian carp: grass carp, silver carp, bighead carp, and black carp. In 2015, the Chinese harvested slightly under fifty million tons of carp. The U, S. Fish and Wildlife Service brought Asian carp (grass carp) to the United States to control aquatic weeds. Bighead carp and silver carp were also acquired by Game and Fish Commissions and put into sewage treatment ponds to reduce the nutrient load by eating algae. Black carp are not nearly as numerous but have also arrived – maybe through the aquarium trade.

Representatives of all four have found their way into assorted waters of the U.S. Grass carp can weigh over 80 pounds, with bighead hitting as much as 100. They can outcompete native fish, becoming the dominant species in rivers and lakes. Black carp feed on mollusks and threaten the survival of some endangered U.S. mollusks. One current effort, concerning these fish, centers on attempting to keep them out of the Great Lakes.

An interesting aspect of silver carp behavior is that they jump from the water, apparently when spooked by a boat motor. They are big fish – up to 90 pounds – and when they are in the air, boats can run under them with boat occupants being struck by the fish, resulting in injuries.

The aquarium trade has populated areas in the Caribbean with lionfish from the Indo-Pacific. These outcompete native fish for territory and food. They are quite colorful, with each in possession of 18 venomous spines. On the plus side, once past the poisonous spines, they are considered good to eat.

Japanese beetles were first found in the U.S. in 1916. In Japan, natural predators keep them in check. In the U.S., in contrast, controlling predators do not exist. The adults eat a wide variety of leaves, and their grubs subsist on grass roots.

It is not just entities that are clearly visible. A fungus entered the U.S. in the early 1900s, and within the next 50 years killed American chestnut trees back to their stumps. While the roots are not dead, any sprouts are typically destroyed by the fungus. These trees comprised about 25% of the hardwoods in eastern deciduous forests. They had been important for both their wood and food for wildlife as well as people.

The previously listed items are a drop in the bucket. We have restructured ecosystems everywhere. There is no natural world beyond strong human influence, and our influence largely consists of homogenization. As an ongoing process, we haul human-associated plants and animals with us as we travel back and forth and around. This process began with our prehuman ancestors. It is a matter of time frame and conjecture as to what is native and what is invasive. There is no time in the past that we can declare to be the natural set-up and aim to restore.

The goal is to structure sustainable systems of plants and animals from which we can feed ourselves in conjunction with a reduced human population. These food-producing systems will remain in flux. Items, based on best guesses and experimentation, should be added and deleted as deemed appropriate and found to be possible. Established groupings would be subject to continuing experimentation and alterations. Perhaps a god did not manufacture us, and this planet and its biosphere with us in mind, but as a softwired species approaching the supraorganic, we, for all practical purposes, own the fucking place and just need to deal with it.

Almost the entirety of the Midwestern prairies has been turned into farmland, with some prairie relics continuing to exist. With integrated permaculture (IP), interested land owners could section off a piece of their property to establish a stable, no-plow area dedicated to long-term harvesting. The first order of business would be to achieve a population of perennial grasses that would lock soil in place and permit its restoration. These grasses would provide, along with their stems and leaves, seeds for animals to eat. Additionally, their seeds could be harvested as are contemporary cereal crops. This is not as big a challenge as one might suppose.

Kernza is a deeply rooted perennial grass that is related to wheat. As a perennial, it can be harvested year after year without additional plowing and planting. The Land Institute, a research organization working with this plant, considers it to have the potential to replace wheat and thus avoid the energy expenditure of annual plowing and planting. Other domesticated cereal crops, such as rye, barley, and oats, would have pre-domesticated variants growing wild in their home range that are perennials. Annuals could be bred to perennial form by taking advantage of rare contemporary variants. Whether pre-domestic variants or newly bred, these long-lived plants would produce deep and permanent root systems.

With the revived grasslands, a population of animals would be installed. Mammals and birds would provide for the harvesting of meat along with the grain. Bison are a logical choice, along with deer and elk. We are not attempting to restore the prairie to some original form, but rather to construct a self-sustaining, extended life-span area from which human food may be extracted. Perhaps some type of large European meat rabbit would fit in. A breed of chicken may also work. Perhaps a type of pheasant could be tried. These projects would be highly experimental and would vary from property to property.

Harvesting animals could be accomplished by means of rounding up larger ones, as with cattle, and smaller ones could be obtained by means of live catch trapping. With reference to harvesting the seeds of grasses, perhaps temporary fencing could isolate conveniently sized areas to keep larger animals out and permit seed heads to develop. These could then be harvested as with contemporary grains.

Integrated permaculture is in no way confined to large land areas as found with farms. It is highly suited to small land holdings and the yards of homes. Michael Judd is a permaculturist and edible landscape designer. He recommends that you replace your lawn with a collection of food-producing plants. Start small with plants producing food that you currently consume – do a garden.

If you have an adequately sized yard, aim for a three-level arrangement (top, middle, and ground). A canopy, or upper level, could consist of apple, pear, or pecan trees. With many trees, it is important to plant more than one to achieve pollination. A second level could contain fig trees or hazel nut bushes and grape vines. A ground level could consist of herbs you cook with. The plantings would need to allow for sufficient light to get through to the third layer. Judd refers to the results as a "homegrown food forest."

The Falkenstein family, in the Phoenix, Arizona, area, grows an assortment of tropical fruit, such as mangoes, sapotes, and a variety of citrus. They have a greenhouse that protects cold-sensitive items during the winter. In addition to plants, they also keep chickens for eggs, meat, and as a source of fertilizer for the plants. They do not eliminate their need for agricultural products, but they do make a significant contribution to both the quantity and the tasty assortment of the foods in their diet.

A restaurant in La Paz, Bolivia, and another in Lima, Peru, serve hunted and fished items from the surrounding rivers and jungle. Mammals, fish, crocodilians, and turtles are brought in and prepared for restaurant customers. This is a case of foraging occurring from a wilderness area convenient to the cities. Farmed items remain crucial to the menu, but hunted, fished, and gathered materials make a significant contribution through the addition of their taste and interesting nature.

If the Mississippi River, from its beginnings to the delta with meanderings and flood plains, were permitted to flow freely, it would provide an enormous source of harvestable food. This would require the abandonment of agricultural lands however, the gain in harvestable food could come close or even surpass that lost through the abandonment of farms.

An estimated 260 species of fish are found across the totality of the river's many branches. Asian carp and tilapia have recently been added to the long-term residents, such as catfish, crappie, and bass. As previously pointed out, the Chinese harvest enormous quantities of carp for human food, while we mostly consider them to be inconvenient and harmful. In addition to the fish, there are an assortment of turtles (such as snapping turtles), crawfish, bull frogs, and various types of mussels that could be eaten. All in all, a fabulous array of items suitable for human food can be taken from Mississippi waters.

The surrounding land presents another bonanza with deer, raccoons, opossums, nutria (another recent addition), bear, rabbits, squirrels, and muskrats. With the possible exception of deer, it is likely that you have not recently eaten many of the other items. Civilized folks consume an increasingly limited number of creatures, and a limited selection of the body parts of those they do eat. Perhaps this is something we need to get over. Not that long ago, we not only ate a bigger variety of animals, but we also ate the entire creature, either directly or made into valued specialties such as sausages and puddings.

Our restricted eating habits exist not only with animals. In the past, more varieties of wheat, other grains, and vegetables were cultivated. With fewer varieties of each item being planted, we are losing genetic diversity. This loss means that our crops are more vulnerable in response to disease or encroaching climate change. Less genetic diversity sets a crop up for contagion. Once again, recall the Irish potato famine. In today's world, genetic diversity is critical for dealing with changing environmental and climatic circumstances.

Items that could be eaten, but are not, frequently become pests: we take things that could be of value and turn them into problems. The nutria population continues to increase in southern swamps. These are mammals that could be of value for both their meat and fur. Nonmigratory Canada geese are pests whose feces can make dog walking unpleasant: they could be Thanksgiving dinner or a treat when company comes. Burmese pythons are 12-foot rib roasts. Seaweeds, such as kelp, can be eaten, used as fertilizer, or processed into carrageenan useful in the manufacture of a wide variety of products.

We may want to eat what we want to eat, but in addition to decreasing the size of our population, in order to fit into and be a part of a long-lasting biosphere, we must expand on our omnivorous proclivities. Learning to harvest, process, and cook a dramatically increasing variety of entities is quite a challenge with considerable potential for pleasure. It is, in reality, a necessity.

Returning to consideration of IP (integrated permaculture) systems, an area in Spain supports a population of pigs feeding on acorns, with the resulting pork supposedly of a superior quality and commanding very high prices. An ecosystem could be constructed in the U.S., possibly in an area of the Appalachian Plateau, with a centerpiece of oak trees and pigs.

Eastern white oaks and burr oaks both produce large acorns with relatively low tannic acid content and represent two possibilities for the trees. Oak trees obviously require decades to grow and begin full-on nut production. Recall that we are escaping from the short-term agricultural practices that have accrued over the last 12,000 years. Also, property could possibly be found with some oak trees already in place.

For the pig part of the deal, a heritage breed such as Tamworth might be given consideration. They would be both sturdy foragers and, although this may or may not be deserving of consideration, more attractive than the standard pink pigs currently raised commercially. The size of the pig population would depend on the available land and would require experimentation. Apparently, in the Spanish free-range pig areas, it runs to about one pig for every six acres. Additional nut and fruit trees would provide more food and reduce the needed acreage per pig.

In addition to providing food for the pigs, various nut and fruit trees, such as English walnut, black walnut, pecan, apple, pear, peach, cherry, plum, persimmon, fig, mulberry, and paw paws could be interspersed among the oaks and potentially provide marketable crops as well as pig food. In addition to trees, bushes such as hazelnut, a variety of grape vines, raspberries, asparagus, Passiflora varieties, and so forth could be added. These items mature much faster than trees and would permit both pigs and an additional assortment of animals to be added more quickly, in addition to possibly providing marketable crops to help fund the project.

As part of a complex ecosystem, other animals would be added to see what would successfully mix in and provide human food along with potential sale items. They could include types of rabbits, woodchucks, chickens, quail, and possibly pheasants. This would be a matter of experimentation over an extended time period. Additionally, at some point, wood would be available for marketing, with many species, such as black walnut and cherry, being of considerable value.

The food available in this system will attract items, such as squirrels and bluejays, that will consume resources intended as food for proprietary animals, or harvesting for human consumption and/or sales. A program could be instituted to trap and kill them, in line with what is permitted by local and state laws. Pigs are omnivorous, and these items could be fed to them, thereby gaining an additional resource.

Eating animals necessitates killing them, and this necessitates consideration as to how this is and may be accomplished. The means by which agriculturally raised cows, pigs, lambs, and chickens are slaughtered leaves a great deal to be desired. Cattle are typically driven into restraining pens and stunned to achieve unconsciousness. The senseless animals are then hoisted by their rear legs and their throats cut to bleed out. For an unknown portion of time, the stunning process fails, and a conscious animal is hoisted and has its throat cut.

Cattle, pigs, and sheep have laws specifying how they are to be killed; chickens, comprising 95% of the animals killed, do not. The standard commercial procedure is for workers to hoist them upside down and place their legs in shackles on a conveyor belt (a process that not uncommonly results in broken and displaced wing and leg bones) that carries them through an electrified bath meant to render them unconscious. The conveyor belt then carries them to an electric saw for their necks to be cut. They are then scalded in hot water for defeathering. Some survive these procedures, and a worker is stationed at the end of the line to cut their throats.

A possible consideration, as a means of killing in a less traumatic manner, is the construction of facilities by means of which the gas nitrogen can be used to accomplish the killing. The air we breathe contains about 78% nitrogen along with oxygen, carbon dioxide, and small quantities of various other gases. Oxygen is, of course, necessary to sustain life. Placing animals in an environment in which oxygen has been adequately eliminated by means of elevating the percent of nitrogen to close to 100, results in animals losing consciousness and dying. Of interest, possibly because we are accustomed to breathing nitrogen, animals do not appear to be traumatized by this means of death. They just lose consciousness and then die.

Consideration of using nitrogen to kill arose with the government animal control services' desire to terminate dogs and cats for which homes could not be found. J. P. Quine describes an experiment in which cats were subjected to breathing a high concentration of nitrogen to the point of losing consciousness, but then permitted to recover from this for two times in a row. On the third time, the cats still did not show stress or attempt to avoid the procedure. Nitrogen hypoxia appears to kill without trauma or distressful sensations.

Animals to be killed commercially could be processed by means of a nitrogen environment rather than being thumped in the head. Apparently, the blow to the head, in addition to being a traumatic means of death, has a failure rate resulting in what can only be considered extreme trauma for the animals involved. With nitrogen hypoxia, the animals pass out without the distress involved with being whacked in the head. Following passing out, they die. If this procedure costs more, eat less meat.

Killing backyard-raised goats or chickens would require a specialized piece of equipment. The results of the animals passing out and dying without the trauma of chopping the chicken's heads off or hauling the goat to an abattoir merit serious consideration. Assorted wild suburban animals, such as raccoons and non-migratory geese, could be trapped and also killed by nitrogen hypoxia, and then eaten. The trauma of trapping could be eliminated by having the animals voluntarily enter a nitrogen device for food contained within it. Studies have produced videos of pigs eating apples within an enclosure and passing out with no apparent stress.

Agriculturally produced food requires the use of trucks, farm machinery, fossil fuels, fertilizers, and possibly herbicides and pesticides. If you grow it yourself, this may be greatly reduced but not totally eliminated. Wild animals feed themselves. Any contribution made to your diet through hunting reduces the size of the footprint left by both yourself and your society and is environmentally beneficial.

The population of persons who hunt, however, is declining. On a positive note, while the loss consists of declining male participation, some replacement is occurring through increasing female participation.

With hunting, we harvest from nonagricultural and nondomestic systems of plants and animals. Wild animals have not been raised by methods that consume fossil fuel or have had hormones and antibiotics emplaced in their bodies. They have not been confined to spaces barely larger than themselves. Nor have they had horns cut off or beaks clipped so that they can be crammed into small spaces and not be capable of injuring each other. They have pursued genetically sponsored behaviors. You be the judge with reference to the morality of hunting in comparison to commercial slaughter.

The discussed projects have all been land-based. Ecosystems just as complex as those could also be put together and managed in either fresh or salt water. Some combination of fish, shellfish, seaweed, and algae could produce products comparable in value to those gained in a new style of prairie.

Saltwater fish farms are in a state of rapid growth, with pollution problems rampant and widespread. With shrimp producing facilities, for example, some are doing a competent job of controlling waste products, while a majority are heavy polluters. It is of utmost importance that these problems be addressed, as seafood is crucial for the feeding of our species, and wild fish populations are being severely depleted.

Commercial saltwater fishing is inadequately managed. Overfishing is rampant with fishing fleets harvesting in an excessive and uncontrolled manner and ranging outside of their allotted waters.. Overfishing not only reduces overall population size, but it also reduces the percentage of breeding size adults, as it is the larger fish that are more frequently caught. This, of course, leads to fewer fish being bred and available for harvest.

New developments are taking place in food production. While greenhouses have been around for some time, new endeavors consist of growing food in warehouses and stacked shipping containers (containers of the type used on ships and trains). In these circumstances, growing trays for planted items are stacked vertically with space between them for grow lights, tubing for water and chemicals, and sensors for measurement of light, temperature, humidity, and plant conditions.

The sealed-off environment of the warehouse or shipping container provides a locked-in environment permitting precise control of growing conditions, including considerable isolation from insects, viruses, and molds. These facilities can also incorporate aquaponics and the use of fish waste as fertilizer. They, furthermore, readily fit into an urban environment with a nearby consumer population or a close by food distribution center.

Very recently, laboratory-grown meat has come into the market. This meat is grown from animal cells with no animals having been killed. The process involves the extraction of cells from animals and then these cells being induced to multiply under laboratory or, eventually, with increasing production, factory conditions. This goes beyond meat-like products produced using vegetable material. This is actual meat grown artificially and not in a cow, lamb, or chicken that would need to be slaughtered to obtain it.

Plants growing in stacked vertical trays under artificial light in a warehouse, and meat growing from animal cells in a factory biodigester, sound like a long way from hunting and gathering. These methods can, however, play a role in feeding ourselves without destroying the environment or driving much of the rest of the biosphere to extinction. With a reduced human population, and feeding ourselves from complex ecosystems and factories that do not pollute and destroy the environment, People Land could successfully exist as a part of the biosphere.

The success of our human population fitting into an ongoing biosphere depends on that population decreasing in size. As previously indicated, our population is predicted to peak and begin to decline by the end of the century. A problem could arise should the human lifespan be extended through medical advances. This has dramatically gained in possibility with artificial intelligence. While the human body is very complex, there is nothing magical about it. Medical personnel, working with artificially intelligent devices, may dramatically extend our knowledge of bodily processes to the point of an extended human lifespan, interfering with population reduction. We may develop an older but still destructively enormous population.

We are out of touch with our environment. We are out of touch with where our food comes from. We are out of touch with the food itself. Vegetables have been cleaned and have undergone various amounts of processing before they are displayed for our purchase. They no longer appear to have been pulled from the soil. The meat products at the supermarket are cut and wrapped. It is easy to avoid the fact that it had once been connected to legs and a head. What we eat appears to have magically appeared and not been grown as part of a plant or an animal.

With catch and release fishing, we separate wild fish from the fish we buy to eat. Catching fish becomes a game in which we drag an animal around by a hook in its mouth and then release it back to the wild. The catching is not one part of a total process, including killing and cooking, by means of which we provide food for ourselves. Fish bought in a store are typically headless and filleted, and just as with pieces of beef, leaving no impression of having once been part of a living thing.

It is not just animals that are divorced from the products we see in stores. Trees are isolated from fruit. Suburban cherry, apple, pear, and other trees, whose ancestors had produced fruit, no longer do so. They have been bred to produce only flowers and not anything to be eaten – they are for decorative purposes only. If they produced fruit, they would be messy and not suited for a suburban yard. On top of that, if they produced fruit, they would attract wild animals. The separation between what we plant in our yards and what we buy at the supermarket further reflects the isolation we have from our food.

Integrated permaculture will play a role in bringing us back in touch with the biosphere that we are a part of and from which we feed ourselves. In Colonial Times, about 90% of the population produced food. An estimate for today is that this is down to somewhere around 2%. Tractors are great for harvesting monoculture. Plucking individual items from a complex mix is more time-consuming and more personal. It is also more emotionally fulfilling, given our hunter/gather evolutionary past.

As artificial intelligence and robotics expand the areas of their competence, fewer and fewer drudgery and unpleasant tasks will require humans to do. You might survive an assembly line or office job, but foraging from a complex, integrated permacultural system might actually be nice. And it should also leave plenty of time for socializing, cooking, eating, and sex.

HUMAN NATURE and the COSMOS

PART II
The Cosmos

CHAPTER 7

Bottleneck
A Difficult-to-Transition Obstruction

My contention is that humanity has entered a bottleneck through which we may not emerge with civilization intact or survive as a species. I did not invent the concept of a bottleneck. Edward O. Wilson in *The Future of Life* labels Chapter 2 "The Bottleneck," in which he gives consideration to the possibility of humanity damaging the environment to the extent of threatening the existence of much of the biosphere, possibly to include our own species.

In this chapter, consideration will be given to four (4) circumstances with the potential to disrupt biospheric processes, leading to the end of civilization as we know it and perhaps the extinction of our species. These concerns are:

- climate change
- nuclear war
- masculinity
- artificial general intelligence

Alteration of biospheric processes, to the point at which the existence of much of life is in question, is a slow strangulation. Even though humanity has been altering the environment over an extended period of time, we have no knowledge of potential tipping points beyond which recovery may not be possible. A life-supporting biosphere may be more fragile than we recognize.

Venus and Earth have a great deal in common: Venus is 7,521 miles in diameter, and Earth is 7,926 miles in diameter. Both are well within our star's habitable zone, in which surface liquid water is at least a possibility. Earth, taking the entire planet and seasonal changes into consideration, has an average surface atmospheric temperature running in the low 60s Fahrenheit. The disconcerting fact is that Venus has an atmospheric surface temperature above the melting point of lead. Just because a planet is 95% the size of Earth and within a star's habitable zone does not mean things cannot be unpleasant.

A strong correlation exists between fossil fuel use and the amount of carbon dioxide in the atmosphere. A further strong correlation exists between the amount of carbon dioxide in the atmosphere and temperature. On a worldwide basis, fossil fuel use is increasing, atmospheric levels of carbon dioxide are rising, and temperatures are rising. Let us pretend it would be more expensive to fix this than it will be to deal with the results. Our great-grandchildren will be impressed with our foresight and wisdom.

On the plus side, renewable energy use is expanding, and the use of coal is stabilizing. While coal use is declining in some nations, India is ramping up use, their perspective being that it is the most efficient means by which the well-being of their large portion of citizens living in abject poverty may be improved. China is putting a great deal of effort into decreasing its dependence on coal, however, it is also building new coal-burning electricity plants - coal is available and cheap. My best guess is that coal use will stay about where it is well into the future.

Even with renewable energy use increasing and coal use somewhere close to stabilized, the U.S. Energy Information Administration projects a 28% increase in worldwide fossil fuel use by 2040, with this increase taking place through the rising use of oil and natural gas. While oil and gas are an improvement over coal, we have a growing world population gaining in affluence and consuming an increasing amount of fossil fuel per person. Atmospheric levels of carbon dioxide are rising and will continue to do so.

Prior to the Industrial Revolution, atmospheric carbon dioxide levels were around 280 parts per million (PPM). We know about this through the analysis of air trapped in glacial ice. In 1958, atmospheric rates of carbon dioxide began to be monitored from Mauna Loa Observatory, Hawaii, and in 1960, they were recorded at 300 PPM. Waldman informs us that PPM rates hit 400 in 2015. Our National Oceanic & Atmospheric Administration (NOAA) claims carbon dioxide rates hit 409 PPM in 2018, 411 in 2019, 417 in 2022, and 419 in 2023. These are the highest atmospheric levels of carbon dioxide in more than 3 million years, and they continue to climb.

Given the preceding, it is a certainty that atmospheric temperatures will continue to rise. NOAA, in 2018, reported that since 1880, Earth's surface temperatures have been increasing and that the rate of increase has nearly doubled since 1975. The warmest 10 years on record have all occurred since 2010. Along with a warming atmosphere, ocean temperatures are also rising. The rate of climb is increasing for both the atmosphere and the ocean, and the effects are becoming more and more visible with planetary ice melting and sea level rising.

Greenland ice melt varies from year to year, with the melt rates from 2017 and 2018 rising, but doing so rather slowly. In contrast, 2019 set a record for ice melt, and 2022 (NOAA data) topped that. Overall, allowing for fluctuations, the melt rate is increasing. The National Academy of Sciences argues that Greenland's ice was melting four times faster in 2012 than it was in 2003. Should the Greenland ice melt continue, with all of Greenland's ice eventually melting, sea level would rise more than 20 feet.

Sea ice surrounding Antarctica varies even more than Greenland ice. The years 2007 to 2016 were characterized by slight increases in the area of water covered by ice. Since 2016, maximum winter ice levels have been decreasing, with the winter of 2023 setting a record for the smallest maximum ice extent. Warming ocean temperatures appear to be playing a role in this. Sea ice surrounding Antarctica is of concern because it hinders the flow of land ice into the ocean. Without sea ice serving as a cork in a bottle, land-based glaciers will flow more quickly into the ocean, playing a role in sea level rise.

Ice reflects more solar energy than open water or bare land. Melting ice exposes water and land that then absorbs more solar energy and further fosters warming. Warmer ocean temperatures, as previously mentioned, increase evaporation and put more water vapor in the atmosphere. Water vapor is a strong greenhouse gas and so traps more solar energy, fostering further warming.

Carbon dioxide and water vapor are not the only greenhouse gases. Methane traps considerably more heat than does carbon dioxide. The warming and melting of arctic tundra results in the release of both carbon dioxide and methane. Even if carbon dioxide accumulation ended, the already melting tundra would continue to release greenhouse gases.

The preceding represents one of many feedback loops. Another being the previously mentioned process of reduced ice exposing more water surface area that absorbs more energy, which then feeds back into increased warmth. Stopping carbon dioxide insertion into the atmosphere will not immediately stop climate change. We cannot make corrections one day and things are OK the next. Just as oil tankers do not steer like sports cars, climate change has momentum.

Climate change is an ongoing natural process that would occur whether or not our species existed. We are, however, having a major impact largely through the use of fossil fuels, both in industry and agriculture. Getting through this bottleneck will require switching to other energy sources. As

Don Huberts, an oil executive, was quoted in the July 1999 issue of *The Economist*: "The stone age didn't end because we ran out of stone." The implication is that the age of fossil fuels will not end because we run out of fossil fuels. We are nowhere even close to doing so.

Enormous quantities of various grades of coal, oil in sand and shale, and natural gas compressed into assorted rock deposits exists. There is no shortage of coal. Oil fields that have in the past been predicted to run out continue to produce as advancing technology permits higher percentages of contained oil to be extracted. And now we have natural gas fracking.

Abandonment of fossil fuels would not immediately fix the carbon dioxide problem because of the sturdy nature of the carbon dioxide molecule. While methane molecules fall apart over around a thirty-year time frame, carbon dioxide molecules last about forever. Carbon dioxide molecules, however, can be sequestered in photosynthetic plant material and the oceans. Burning fossil fuels releases carbon that photosynthesis had stored.

Oceans both absorb and release carbon dioxide. With global warming, and the oceans becoming warmer, they absorb less. The carbon dioxide that is absorbed reacts with water to form carbonic acid, raising the acidic level of the oceans and reducing the ability of organisms, such as clams and corals, to form and maintain shells. Overall, ocean currents and natural processes affect the intake and release of carbon dioxide. Oceans do not make it just disappear.

Farming, with its plowing, harvesting, and soil erosion, provides for minimal photosynthetic carbon storage. Complex, long-term ecosystems with roots deep into the soil, as were promoted in the preceding chapter, would store more carbon. This further emphasizes the value found in rejection of the agricultural paradigm and the practices it promotes. Integrated permaculture contributes to a healthy biosphere.

Current levels of environmental damage are already impacting lives. If you have coastal property, it may previously have taken a storm to produce flooding, whereas currently, flooding may occur simply with a high tide. For inland residents, an increasing percentage of warm days jacks up energy bills. Then there is the matter of water for gardens and lawns. The changes that have already taken place are not inconsequential, and as climate modifications accelerate, the problems will accrue and grow in consequences.

An increasing and increasingly affluent world population requires more food from crops suited for climates that are shifting. Temperatures and rainfall change. We lose land with rising sea levels. A warmer ocean means more energy available for storms and an increased evaporation rate. Higher levels of atmospheric moisture lead to more rain, not necessarily where it has been falling or where it would be beneficial. Keep in mind, these changes are ramping up.

We probably will not achieve the climate of Venus, but nothing is a certainty. Earth has experienced considerable climate variations in the past, with these presenting possible parameters for the future. Our planet has experienced times known as "Snowball Earth" during which it was almost completely frozen over, as well as times with no ice at all. And, there may be tipping points beyond which Earth could achieve new extremes. Should current circumstances proceed uninterrupted, we will almost certainly achieve another "no ice" situation with a sea level rise of over 200 feet – there goes 9/10th of Florida, along with a great deal of the East and West coasts and their cities. Should current circumstances continue unabated, this level of flooding is not a matter of "if" but of "when."

Climate change as a means of biospheric destruction is time-consuming. Nuclear war is a quicker and more efficient suicide. Nuclear war, on its face, openly threatens human extinction. Wilson refers to war as "humanity's hereditary curse." With the availability of nuclear weapons, a failure to exorcise this curse sponsors the end of life as we know it and possibly the extinction of our species. Edward O. Wilson in *The Social Conquest of Earth*, refers to our species as having an "innate pugnacity" "ingrained because group-versus-group was a principle driving force that made us what we are." Nuclear weapons are a game-changer. Our propensities for intergroup violence transit from competition to self-destruction.

Along with deaths from explosions, read John Hersey's book *Hiroshima* for a look at the consequences of radiation. A nuclear war would have multiple areas of destruction surrounded by worldwide debilitating and killing radiation. Additionally, it would loft enough particulate matter and smoke into the atmosphere to bring about a nuclear winter.

The term "nuclear winter" was coined by James Pollack, Brian Toon, Thomas Ackerman, and Richard Turco, whose interests span dust storms on Mars and the effects on Earth's atmosphere of the asteroid impact of 66 million years ago that resulted in the Cretaceous-Tertiary (K-T) boundary and extinction event.

With the K-T extinction occurrence, around 70% of all animals became extinct, including mammals, birds, reptiles, and fish; it was not just the famous dinosaurs, but mostly everything bigger than a house cat, along with much else. Elizabeth Kolbert, in her work, *The Sixth Extinction: An Unnatural History*, provides an excellent overview of the K-T event along with a thoroughgoing presentation of the extinction episode our species is currently orchestrating.

Luis and Walter Alvarez connected the K-T extinctions with an asteroid impact. The Alvarezes believed it was not the impact itself that caused widespread extinctions, but the resultant material lofted into the atmosphere. This was right up Pollack and company's alley. Atmospheric and land impact of the bolide would have generated enormous heat in addition to blasting material into the air that would have spread around the world and incandesced as it fell. Heat from the blast, along with material aflame with falling through the atmosphere, would have generated worldwide fires, adding smoke to already present atmospheric contamination.

The lofted material and smoke would have served as a shield reflecting sunlight back into space. A lack of incoming solar energy would have inhibited photosynthesis and, with reduced solar energy reaching the land, resulted in lower temperatures. Less solar energy and colder temperatures would have produced a dramatic reduction in the vegetation that all animals ultimately depend on. The impact itself would have ended life in an enormous area, but the extinctions over the entire planet provide evidence for atmospheric involvement.

The K-T extinction event was brought about by the impact of a single bolide. The coiners of nuclear winter have advanced the idea that the multiple detonations to be expected, should major world powers enter into nuclear conflict, would also do the job. Large quantities of debris and dust, along with dust from burning cities and petrochemical fires, would enter the atmosphere and spread around the world. Just as with the asteroid impact, this material would block sunlight, resulting in diminished photosynthesis and falling temperatures. The end result would be widespread famine arising from severe hampering of agricultural production on a worldwide basis.

A nuclear war could reasonably result in an extinction event comparable to the K-T incident. Even if our species did not become extinct, civilization as we know it would no longer exist. Should you reject the concept of nuclear winter as leftwing rubbish, immediate destruction from the detonations and long-lived despoliation from globe-circling radiation are widely acknowledged realities. It would be/lose-lose circumstance for participants and the audience alike. Surely our "wise and thinking" species would not enact a nuclear World War III.

Unfortunately, we are not actually wise and thinking. We are a species of advanced sociality approaching the supraorganic. Evolutionarily programmed emotion, and not rational thought, rules relations between both individuals and nations. Assumptions of rationality led to the belief that World War I was so terrible that it would end all war. If we were actually rational, it might have. World War II saw the detonation of two nuclear devices. Regardless of the horror they brought about, major world powers have loaded up on more advanced and destructive versions. If human impact on climate is a problem, it is a piece of cake compared to this.

It is a good idea to look at what people do and not what they say. Written history and archaeologically explored prehistory present a record of unending war. Recent United States history supports this assertion. The following is a list of wars and military interventions that have been fought during my lifetime:

World War II	1941 - 1945
Korean War	1950 - 1953 (no official end)
Vietnam	1961 – 1973
Dominican Republic Invasion	1965
Lebanese Intervention	1982 – 1984
Grenada Invasion	1983 – 1985
Panama Intervention	1989
Gulf War	1991
Somalia Intervention	1993

Haiti Intervention	1994
Bosnian War	1994 – 1995
Kosovo Intervention	1999
Afghanistan War	2001 – 2021
Iraq War	2003 - 2010
Russian/Ukrainian War	2022 – present. We are currently supplying equipment, weapons, and ammunition to Ukraine
Israel/Hamas War	2023 – Begun as this was being written - the U.S. is a close ally of Israel

Wars are ongoing, with nothing indicating that they will ever cease to happen. When the United States breaks stuff and kills people in less technologically developed areas, life goes on for its own citizens. It is not that this sort of thing cannot get expensive or generate difficulties at home, but it is a different game when world powers are fighting among themselves as they were in World War II. This is the worrisome part of the United States' support for Ukraine. Will we be able to maintain our proxy status as a supporter of Ukraine, or will we get sucked into war directly with Russia?

Should the United States and Russia confront each other militarily, without an intermediary, both would have their survival at stake. Under these circumstances, nothing is held back: nuclear weapons would be employed. War is an aspect of the United States' foreign policy. The big question concerns whether or not we will manage to confine ourselves to supporting the conflicts of others and avoid confrontation with another nuclear state.

In addition to the hostile nature of the US/ Russian relationship, China would like to replace the United States in its perceived position as the leading global power. Both are driven by imperial aspirations, currently in collision in the Asian Pacific. The United States has been dominant in this region since World War II and strives to maintain that position. China, a resident of the Asian Pacific, wants to usurp that dominance, and ultimately the dominance held by the U.S. on a worldwide basis. An ongoing series of confrontations and clashes is assured.

Both the United States and China are the equivalent of schoolyard bullies. When the Vietnamese kicked out the French, the United States moved in and assumed the mantle of French colonialism. The reason given was that the newly empowered Vietnamese government was leaning in the direction of communism. Communism, it seems, needed to be stopped in Vietnam or it would spread like a cancer and threaten democracy and the way of life in the U.S. It took 12 bloody years for the Vietnamese to throw out the Americans as they had thrown out the French.

The United States, in conducting the war, bombed not only Vietnam but also surrounding countries, including Laos and Cambodia. In Laos, the U.S. "dropped more than two million tons of bombs." This is "more than it dropped on Germany and Japan during World War II." In addition to slaughter at the time of the bombing, these countries now have a sprinkling of unexploded ordnance that continues to maim and kill. Additionally, the air war succeeded in destabilizing the region and, thereby, played a role in the Cambodian genocide that followed.

Since Vietnam, the United States has continued to maintain an "I'm the boss" attitude, managing to offend just about every nation in the region. The Philippine government, in 1992, became so put out that they forced the closure of the United States naval base at Subic Bay. This was no small deal for either the United States or the Philippines. For the U.S., it meant loss of a major base from which power could be projected. For the Philippines, it was a loss of insularity with China.

Discord between Pacific nations and the U.S. continues. In 2018, eleven Pacific nations signed the Trans-Pacific Partnership (TPP) trade deal aimed at reducing tariffs and promoting trade. The U.S. failed to participate based on the belief that, even though the agreement would benefit the region, it did not adequately serve the United States' interests.

Contention is not limited to that between the U.S. and Asian nations. A conflict concerns the United States, Canada, and the European Union's use of Pacific nations as convenient places to dispose of recyclable scrap such as plastic and cardboard. Malay's environmental minister, Yeo Bee Yin, criticized the flow of imputed recyclables as, in reality, garbage being dumped in Asia. Rodrigo Duterte, the then-president of the Philippines and a more colorful personage than Ms. Yin, threatened war with Canada if they did not take back a shipment of their garbage.

While the United States is on the other side of the planet, the Western Pacific is China's home. Nations in the Chinese neighborhood generally may go their own way as long as they accede to Chinese dominance. This dominance includes unreasonable claims to ocean sovereignty and fishing rights.

Vietnam, following the Vietnam War with the United States, annoyed China by invading Cambodia. China's response was to invade, but not occupy, Vietnamese territory. China does occupy Tibet, and considers Tibet, along with Manchuria, to be part of China. Also, with reference to possible future neighborly activities, China considers a chunk of Eastern Russia to more appropriately belong to China. Whether occupying or just being neighbors, both China and the U.S. have an "I'm the boss" attitude.

As a clear indication that Chinese ambitions extend beyond its own neighborhood, and are global and imperial in nature. China is expanding its navy and has established its first overseas military base in Djibouti. Aircraft carriers are among the ships being added. These boats, as part of battle groups in conjunction with other war and supply ships, are de rigueur for the projection of naval power on a worldwide basis.

Regional Chinese dominance is expressed by assertions of territoriality and sovereignty over areas of ocean in some cases also claimed by the Philippines, Viet Nam, and Taiwan. The layout of national coastlines in South East Asia sponsors conflicting claims of ownership and fishing rights. China participates in these disputes by gobbling up enormous areas for itself. China considers these gobbled areas to be sovereign Chinese territory over which it controls not only fishing but also ship traffic.

In justification of its claims to oceanic water, China constructs islands by dredging material onto rocks and reefs. In addition to the newly manufactured islands rationalizing this sovereignty, they serve for the construction of military facilities and airfields. The islands then act as a pretense for a claim of ownership to surrounding waters, as well as a defensive outer perimeter to protect the mainland, and as bases from which power can be projected.

Confrontations between the U.S. and China occur as a result of the U.S. not recognizing all of China's sovereignty claims for expanses of ocean. The U.S. considers some of the areas over which China claims sovereignty to be international water open to all ship traffic. The U.S. enforces this contention by transiting naval vessels through the areas in question. The Chinese response is bellicose anger. On occasion, ships from the two countries cruise toward each other, like teenage drivers playing chicken, barely avoiding collisions that would take lives and possibly start a war. This behavior should dispel any consideration being given to the possibility of the United States and China constituting sapient, rational players, or generally behaving in a manner we fanaticize as appropriate for adults. World War III? What me worry?

It is not just the U.S. and China to be worried about. India and Pakistan both have nuclear arsenals and occasionally shoot at each other. The Middle East is always on the verge of war, somewhere. The U.S. has imposed financially damaging sanctions on Iran in response to its attempts to join the Nuclear Club, leading to retaliatory Iranian activities. The Kurds have no homeland. The Russians and NATO are generally in a snit about something: today it is the Russian/Ukraine war. Many nations have internal groups vying for power and willing to shoot it out. There are never ending, ongoing, violent confrontations sprinkled about on this planet. Will, or perhaps more realistically when, will one of them blossom into the next major war?

A comforting thought is that war is an aspect of contemporary social structure and if we go back to prehistoric times, the world was considerably more peaceful. If this were true, and war was a rather recent phenomenon, it should be amenable to extirpation. Unfortunately, what the archaeologist Lawrence Keeley, in his work *War Before Civilization: The Myth of the Peaceful Savage*, points out is that wherever he has dug, evidence of violence was found, and frequently violence extensive enough to reasonably refer to it as war. He states that war is "universally common and usual." The commonality of war in my lifetime (the episodes previously listed in this chapter) argues in support of Keeley's position and of war as standard societal practice. It has been this way back through history and prehistory.

Keeley further points out that "raids and ambushes are the most frequent and widely employed form of nonstate warfare." This behavior consists of members of one community attacking the camp or village of another community, or surprising and killing an individual or members of a group from another community. This is very similar to chimpanzee incursions into neighboring territory, attacking and fatally injuring isolated individuals. The difference: human groups do not just attack individuals, but go after groups and villages. Chimpanzees attack with body parts, such as hands and teeth, inflicting wounds that typically prove fatal following the incident. Humans fight with weapons capable of immediate lethality. Human violence does not just go back into our prehistoric past, it goes back into our prehuman past.

If our species has engaged in a behavior, going back without interruption into our prehuman past, if that behavior is currently underway at multiple locations, and if no profound developments have occurred regarding this behavior, it is a good bet that the behavior will continue. The enormous effort and money contemporary nations are putting into preparations for war show a profound faith in its continuation. The level of effort further enhances the probability that major conflicts will be sprinkled about along with the never-ending smaller ones.

Actually, there has been a profound development. The last major war saw the invention and use of nuclear explosive devices. With this knowledge and technological capability, war went from merely destructive and murderous to shoving us into a bottleneck from which we may not emerge. If war is a certainty, is it now also a certainty that major conflicts will, at some point or another, involve nuclear weapons?

We live in a chemical world. When looking around, it becomes obvious something else is out there. Some minerals are what is referred to as radioactive. They exhibit energy apparently not originating from chemical processes. Something is going on with them that is different from when coal burns. Also, as we gained knowledge concerning how old things are it became obvious the Sun must be something other than similar to a burning lump of coal. If it were powered by molecular processes it could not have existed for anywhere nearly as long as it has.

It became obvious that there was a source of energy out there beyond the chemical. Maybe it could be employed in war? No one knew. The U.S., Germany, and Russia worked on it. The U.S. got there first and used the new bomb on Japan in 1945, thereby ending WW II and passing through the doorway into a nuclear world.

Now everyone knew an atomic bomb was possible, and the race was on. Russia made a big surprise by detonating a bomb in 1949. Britain followed with a detonation in 1952, France in 1960, and China in 1964. Israel apparently has nuclear devices, but has not detonated any, and at least to some extent promotes the pretense of not having them.

A great deal of effort has gone into limiting membership in the Nuclear Club. Following the destruction of Hiroshima and Nagasaki, the U.S. had aspirations of keeping the knowledge necessary to construct a nuclear device a secret. The Russian detonation, a couple of years following Hiroshima, exhibited the flaw in the plan: nuclear processes are a part of the natural world and available to any nation investing in appropriate scientific research, along with the standard practices of spying, theft, and purchase.

India and Pakistan joined the Nuclear Club, not because they were invited but because the old members were unable to keep them out. North Korea and Iran are currently working on becoming members, but have not yet succeeded in loading up on lethal instruments, with the current members doing everything manageable to keep them from doing so. Regardless of efforts to deny membership, the Club will grow. Any technologically advanced nation can go nuclear. With war, given human nature, a never ending and inevitable social process, the more nations in possession of nuclear weapons, the greater the certainty they will be employed.

Perhaps Nuclear Club members will disassemble their stockpiles of bombs and we can back out of the atomic part of the bottleneck we have entered. Unfortunately, those currently in possession of assorted atomic devices exhibit zero actual intentions of giving them up. In 2003, U.S. National Security Advisor Condoleezza Rice, in reference to North Korean ambitions concerning nuclear weapons and in total disregard of her own nations collection, stated that North Korea needed to "recognize that if they are going to flaunt their international obligations, there would be a cost for it." If we do it, it is an understandable necessity. If you do the same thing, you are flaunting international obligations. It is an interesting display of "do as I say and not as I do" deportment that proves not to work any better in international relations than it does in parenting.

The U.S. and Russia both make noise concerning the reduction in their stockpiles of nuclear weapons. Their combined inventory peaked in 1986 at over 60,000 devices. This is now down to under 10,000. While this is a nice reduction, those remaining would reasonably be adequate to initiate a nuclear winter.

The real reason for the reductions were that both nations simply had grossly excessive quantities, well beyond anything that could be of value, and all retained devices would need to be modernized at considerable expense. With reference to modernization, the March 7th, 2015, issue of *The Economist* asserts, "every nuclear power is spending lavishly to upgrade its atomic arsenal." Presenting reductions in inventory by Russia and the U.S. as arms control is a farce. Reductions were actually about economic necessity. While Russia and the U.S. are putting on a sardonic comedy extravaganza, China is simply stocking up.

The idea that nuclear armed nations, having attained their nuclear capabilities at great effort and expense, would not use them if faced with their society's destruction is unrealistic. We are the same species of ape that we were before the acquisition of nuclear weapons. We have not suddenly become sensible. And, we prepare for war.

Regardless of the horrendous aspects and consequences of war, nations continue to prepare for it, thereby making it a consideration. China upped its military budget by 83% between 2009 and 2018. A difference between the Chinese military budget and the U.S. military budget, in 2018, is that even with the enormous increase in Chinese spending, the U.S. still spends over three and a half times more than China.

When circumstances exist in which the desires of one nation are incompatible with the desires of another, they will both see their own wishes as just and reasonable and those of the other as grasping and wrong. They will see themselves as rightfully deserving of whatever is desired. It has been this way since we were the ancestral species of ourselves and our ape cousins. We fail to see our immense human commonalities in the face of conflicting desires. We see ourselves as the good guys and the others as orcs. The United States and China both want to be the big dogs in the Pacific and the world. It just seems right to both of them.

As this is being written, Russia and Ukraine are fighting a war. Russia believes, based on its historical perspective, that Ukrainian territory should actually be a part of Russia. Iran apparently agrees with Russia as it is supporting their war efforts. Much of Europe and the U, S. are supporting Ukrainian war efforts. Under these circumstances, countries cluster in groups supporting the nation they feel to be in the right. These assemblages set the world up for a World War III with Russia and the U.S. being on opposite sides and both having nuclear weapons.

Masculinity is given separate consideration because it is the driver of war. Males were, as previously pointed out, evolved to specialize in securing and defending territory essential for community wellbeing and survival. The males of a community engage violently with neighboring communities. A community whose males did not do so would have lost its land.

Male-sponsored intercommunity relations are more complicated than just violent interactions. Males are also involved in forming alliances between individual communities that serve to unite hostile efforts toward other groups. Loaded in with all the violence are trading relationships. The males establish extensive trading patterns that move items over long distances through multiple communities. The complexity of human community relationships distinguishes them from those of chimpanzees, who only interact with physically adjacent neighbors and do not engage in alliances and trading.

The violence of war is necessitated and embedded within a generalized male violence. It is a package deal. To avoid the self-annihilation of nuclear war, the overall, evolved, testosterone-fueled, masculinity-fabricated totality of male violence must be dealt with. Violence internal to European, British, and North American societies has been successfully reduced through the management of societal input supporting violence. However, it appears that these efforts are not adequate to end male commitment to war.

War continues to threaten the survival of human civilizations, and possibly the extinction of our species. These circumstances justify considering radical solutions. Lionel Tiger, in his work *The Decline of Males*, points out that barbiturates and other medicines taken by a pregnant woman can impact fetal testosterone production and thereby influence later adult male behavior. He refers us to the work of June Rheinisch and Stephanie Sanders, researchers with the Kinsey Institute, who have found that medicines taken by pregnant women affect the adult personalities of their offspring.

Research could determine what medicines a pregnant woman could take that would influence a male fetus to lessen the masculinity of the resultant adult. While this may be well beyond the pale in today's world, what remains of humanity following a world nuclear conflagration may be more open to such extremes.

In addition to all of the preceding, we now have artificial general intelligence (AGI) looming on the horizon. A great deal of money and effort is going into achieving advances in AGI. Existing entities can write sonatas and dissertations. At some point, it will achieve parity with human intelligence and cognitive abilities, and then surpass them – it is a matter of when, not if.

In counterpoint to efforts and advancements in AGI, multiple sources assure us that artificial devices will never fully be our equal. As an example of this, the columnist David Brooks, who writes for *The New York Times*, stated that "AI will force us humans to double down on those talents and skills that only humans possess." That there exists a "humanistic core" within us that a non-human device can never replicate. That there is "knowledge that is useful information," and there is "humanistic knowledge that leaves people wiser and transformed." Perhaps we are really and truly just so special, and perhaps this is delusional and wishful thinking.

Contemporary human intelligence and cognitive abilities do not represent any sort of upper limit as to what is possible. It is what we have achieved through evolution. There is no reason to assume that this could not also be fabricated. There is no reason to assume that something made out of plastic and metal could not equal or surpass something composed of meat. There is nothing magic about meat.

Should artificial devices achieve levels of cognition beyond that of our species, it may be foolish to assume that life would just go on as it always has. Independence for machines will arrive with intelligence beyond that of humans. Following this, our options may be to achieve some degree of artificiality ourselves in order to keep up with intellectual machine developments, or accept a status of being to them what hamsters are to us.

A difference, and almost certainly an important one, is that the artificial entities will not have had millions of years of evolution with evolved brain parts lying behind their cognition. They may or may not place the same value on individual unit survival, or staying alive, as humans do. They may or may not put efforts into furthering their intellectual capabilities. And, perhaps most importantly of all, they may or may not concern themselves with the biosphere and its biology.

Artificial devices with high levels of general intelligence, beyond that of our own, may take an interest in our species, as we are the most complex creatures around. As such, they may simply find us interesting and observe human proclivities, or they might try to preserve our existence and civilizations by interfering with self-destructive behaviors such as our potential for nuclear war.

It could also be that their interests would lie with the total complexity of the biosphere and see our species as the most dysfunctional and damaging part of it. As such, they could conceivably decide that the best thing for the total ball of wax would be to exterminate us.

Surviving our bottleneck may or may not be a possibility. A good bet is that it will not be easy. Are we alone with this conundrum, or is it something that civilizations on other worlds frequently face? What circumstances do we share with other worlds? The following chapter gives consideration to our commonalities and embeddedness within the Cosmos.

CHAPTER 8

Imperatives
The Same Everywhere

No matter where you go, chemistry and physics will always be the same. This is also true of evolution. No matter where you go, given appropriate circumstances (whatever they are), living things (whatever they are) will come into existence and evolve with sociality as a theme. Even though a lot of things will continue to survive with minimal levels of interaction, cooperation will provide a competitive advantage to those that pursue it. Complex cells will evolve, multicellular things will evolve, and the multicellular things will cooperate to the point of becoming suprathings.

There will always be, as on this planet, two ways to achieve complex social structure: one is a hardwired and chemical approach, while the other is a softwired method in which some form of inherited behavioral propensities mesh with social interactions. Hardwired and softwired are like poetry and prose. At times, they might get muddled together, but everything fits in somewhere. On this planet, the hardwired approach has resulted in fully supraorganic ants, with individuals no longer being individuals but serving as dependent, specialized components of a larger entity. Our species, employing a softwired approach, has moved strongly in the direction of, but not fully arrived at, supraorganic status. At various levels of social organization – simple cell, complex cell, multicell, supramulti-cell – clear benefits are demonstrated arising from the pursuit of social complexity.

Alien lifeforms almost certainly will not look like ants or people, but given enough time, there will be a hardwired version of something pursuing a lifestyle similar to ants, and something else pursuing a softwired lifestyle similar to people: an alien version of People Land. The supraorganic ant approach is so successful that when researchers collected and weighed all the animals in a single hectare of Amazonian rainforest, "ants alone weighted four times more than all the terrestrial vertebrates – that is, mammals, birds, reptiles, and amphibians combined." In terms of mass as an indication of species success, all ants living on Earth weigh roughly as much as all the humans.

A softwired lifeform with multi-generational communities that deal with their environment as a unified whole, as our species does, will evolve. Both they and we may or may not achieve full supra status, but human communities do come close with dependent members having socially transmitted specializations and working together to deal with the environment as a unit. The key here is "socially transmitted." Pheromones control the hardwired world, while the softwired depends on socialization. Whereas the hardwired modulates behavior chemically, in the softwired, individuals read emotions, facial expressions, and vocalizations as determinants of interactions, empathy rather than pheromones. The big thing missing for humans is to have specialized members do reproduction rather than everyone just having at it.

Hardwiring evolves and appears much faster than softwiring, but once softwiring arrives, it then evolves much faster than hardwiring. "The eusocial insects are almost unimaginably older than human beings. Ants, along with their wood-eating equivalents, the termites, originated near the middle of the Age of Reptiles, more than 120 million years ago. The first hominins, with organized societies and altruistic division of labor among collateral relatives and allies, appeared at best 3 million years ago." Admittedly, we are not fully supra.

Going back to the basics of the plants and animals of this planet, alien stationary items will evolve along with mobile forms that wander around and snack on them. The next big universal opportunity is for the evolution of something that snacks on the snackers. There are just so many ways to make a living, leading to convergent evolution. With reference to the snackers, both Australian marsupial mammals and North American placental mammals evidence a cat and a dog approach. In Australia, the carnivorous spotted-tailed quoll follows the cat approach and ambushes prey at night. The now-extinct thylacine or Tasmanian tiger (the last one died in captivity in 1936, with continued questionable reports of sightings in the wild) pursued a carnivorous dog-like lifestyle characterized by stamina rather than speed, in which prey were pursued to exhaustion.

Regardless of where life evolves, commonalities will exist. Just as we have grass and grazers on Earth, alien versions of low-growing plants will exist on planets in conjunction with alien creatures that eat them. The snacking things then provide an opportunity for alien versions of carnivores that can specialize in either speed or endurance. Just as with chemistry and physics, evolutionary principles will be the same everywhere, and there will be commonalities based on possible ways of making a living. The preceding arises from our planetary perspective. Additional complexity appears when we go beyond that.

We only know of life on the surface of Earth. Life could also possibly evolve in the upper atmosphere of a gas giant planet or in a liquid found on something such as a moon of Saturn. It might not be carbon-based or have anything resembling DNA. As a commonality of what we are familiar with, it will, of necessity, need to make a living. My guess is that it would need to consume something to provide energy for the maintenance of its existence and potential reproduction. With reference to sociality, it could exist on any level from isolated individual units to the advanced sociality of something with which we could communicate.

Should an alien version of life pursue sociality, it would, as we did, go through a phase in which it mistakes advanced sociality for intelligence. David Grinspoon used the term "proto" to describe humanity's current state of affairs with reference to intelligence. He goes on to optimistically raise the possibility that proto may be a phase preceding actual intelligence. Grinspoon further raises the possibility that the "proto-intelligent phase may be one that most species don't make it through." The idea of a bottleneck certainly does not originate with me.

The pursuit of sociality is a universal evolutionary theme, and all developing, human and non-human, civilizations mistake advantages arising by means of it as arising due to their brilliance. Under these circumstances, as on Earth, the pursuit of perceived interests would lead to serious damage to the biosphere and drive a great deal of other life forms to extinction. Additionally, as with our species, the desires of an individual's community would be perceived as just, while identical desires of the neighbors would be perceived as greedy and mendacious, leading to unending wars. Nuclear processes and energy are an aspect of the natural world amenable to discovery and use in weapons. They would enter the same bottleneck as we have, and who knows who will or who will not survive.

The preceding ideas are, of course, speculation. We have no direct knowledge of the existence of life anywhere other than on this planet. A base assumption being made is that the Earth is not an exceptional place and that chemistry, physics, and evolutionary processes are the same everywhere. We are reasonably certain concerning chemistry and physics, but whether or not life even evolved anywhere else, let alone followed identical evolutionary processes, is up in the air. We do not have a clue as to how frequently life actually comes into existence in off-Earth environments.

Support for the idea that life is not rare comes from the speed at which it appeared following the diminishing of the bombardment of material forming the Earth. David Grinspoon states, "the planet became inhabited as soon as it was habitable." When the accumulation of material from space lessened enough to permit life to form, it did so. That it appeared so quickly on Earth supports the contention that it would also readily appear in other places.

As Edward O. Wilson points out with reference to Earth, "virtually every square inch of the planetary surface" is inhabited by "creatures of one kind or another." If Earth is a normal sort of place, and life quickly appeared here and then spread widely on its surface, this supports the contention that life will appear in off-planet places and spread. And, even if Earth is a bit unusual and the appearance and spread of life is not all that common, there are so many other places it still would not necessarily be all that rare.

There are uncountable stars out there with planets proving to be standard accompaniments. We know life has happened at least once (look in a mirror). Even if life-producing circumstances are quite rare, the principle of plenitude asserts that with stars existing in numbers beyond human comprehension, and with planetary systems as normal accompaniments, it will exist in other places.

If life does exist somewhere out there, the next question concerns how frequently it will result in something like us. Two extreme views are that our existence depends on a long series of lucky events that make human existence highly unlikely; with the opposing view presented here being that something going in the direction of humans is almost a certainty given the evolutionary theme of sociality. Edward O. Wilson presents the view that we exist based on a series of improbable events, all of which must have occurred in sequence for us to exist. "The reason is simply the extreme improbability of the preadaptations necessary for it to occur at all."

Matthew Cobb, a professor of zoology at the University of Manchester, states an argument similar to Wilson's. In his chapter in Al-Khalili's book *Aliens*, he argues that the evolution of complex cells, followed by multicellularity, followed by "self-awareness and civilization," are each an improbable event. "Multiplying these improbabilities strongly suggests that we may indeed be alone." Each of these improbable events must occur sequentially for something on the order of us to exist.

Cobb further argues that if the bolide strike of 66 million years ago had not occurred, reptiles would have continued their dominance and no niche would have come into existence in which mammals, and therefore we, could have evolved in. He goes on to state: "we cannot assume that some kind of highly intelligent reptile would have evolved in our place. There is no evolutionary tendency for animals to become more intelligent, or more complex." He is wrong. It is not intelligence. It is sociality and it is an evolutionary theme. Complexity happens.

They might not look like humans with odd bumps on their heads as the alien lifeforms do in Star Trek, but if my ideas concerning sociality being an evolutionary theme are in touch with reality, they will have experienced the same bottleneck that we are currently experiencing. I use the past tense for them and the present tense for us with my assumption being that for them to show up here, as a spacefaring civilization, they could reasonably be expected to have survived and gone beyond their bottleneck. In order for us to go wherever they are, we will necessarily need to successfully transit our own. Meeting the neighbors would mean that at least one lifeform has made it through. Maybe most do, or practically none.

Most of us want to know if other life exists in the cosmos and especially if it happens to be something with which we could converse. The first and most basic step concerning discovering whether or not life evolved elsewhere was to determine if other planets existed on which it could come into existence. As previously mentioned, a concern specifically with planets arises from our species-circumscribed experience. Given our limited perspective, planets appear to be essential, and planets orbiting other stars began to be discovered in the 1990s. Currently, thousands of extrasolar planets have been found. We now know planets are a normal accompaniment of stars.

One method of planetary discovery is to look for a decrease in a star's brightness due to a planet crossing in front of it. This method requires the orbit to result in the planet passing in front of the star with reference to our line-of-sight. A second method is based on orbits occurring around the center of mass. A star does not sit stationary while a planet moves around it. The star also moves in response to the mass of the planet. They both move with reference to their center of mass. This means planets can be detected based on the wobble of their stars. There is no shortage of planets.

The next question concerns the existence of planets adequate for the evolution of life, with the search being structured around what we know from our home planet and none other. Given our very limited perspective, it appears liquid water is essential. Who knows? Maybe life exists in the atmosphere of Jupiter.

We have discovered a great many large planets because they dim stellar light output more than small ones, and because they make their star wobble more as they orbit because of their mass. Current instrumentation is just barely able to look for rocky planets, somewhat similar to Earth in size, and located in the Goldilocks zone in which liquid surface water is at least a possibility. The Goldilocks zone is dependent on a planet having an orbit that is far enough away from its star so as not to have all of its water boiled away, and yet close enough for that water to not remain constantly frozen.

Whether or not a planet is in the Goldilocks zone is subject to a SWAG (Scientific Wild Ass Guess). A guess as to the existence of liquid surface water remains iffy both due to the current levels of our instrumentation and the complexity of planetary processes – Venus, as with Earth, lies within the Goldilocks zone of our star, and yet, as previously indicated, has a surface temperature above the melting point of lead. The procedure for making that guess consists of spectrographic analysis of light from its star passing through the planet's atmosphere. In addition to our technology being minimally up to the task, our position and the position of the planet and its star must be perfectly aligned.

Working with images taken by the Hubble Space Telescope, in 2019, Angelos Tsiaras, an astronomer with University College London, studied light filtered through the atmosphere of a planet, designated K2-18b, circling the star 51 Pegasi. K2-18b appears to be a rocky planet about eight times the size of Earth, located in the best guess as to the star's Goldilocks zone. Most strikingly important is that spectroscopic analysis indicated water vapor in its atmosphere, thereby establishing it as potentially life-supporting.

Two newly orbiting telescopes, dedicated to finding Earth-size exoplanets and providing spectrographic analysis of their atmospheres, are dramatically increasing available data. National Aeronautics and Space Administration's Transiting Exoplanet Survey Satellite (TESS) and European Space Agency's CHaracterizing ExOPlanet Satellite (CHEOPS) are both dedicated to finding and analyzing the atmospheres of small rocky planets. We will be learning a lot more about potentially life-harboring planets.

If a planet is in what appears to be its star's Goldilocks zone, and its atmosphere contains water vapor, it is a candidate for a biosphere. If life has actually evolved, further spectrographic analysis of the atmosphere may reveal evidence of its existence. The TESS and CHEOPS satellites could make this possible. On Earth, for example, life has added oxygen to the atmosphere. For the atmosphere to contain oxygen, it must be continuously added because oxygen combines with other elements, such as iron, thereby decreasing atmospheric amounts. An unstable atmosphere requiring constant input is indicative of something going on. It might be some geological process, or it could mean life.

Keep in mind that we are not just looking for bugs, but something with which we might converse, take drugs, have sex, and establish relationships: just seeking out biospheres is not enough; we want to find civilizations. One argument is that civilizations progress through phases in which they harvest and use an ever-increasing amount of their star's energy. They might possibly use material from an asteroid belt to build solar energy collectors across vast distances within their planetary system. We would just stumble upon this sort of thing as we perused the skies. The underlying philosophical perspective being that true meaning in life can be achieved through acceptance of the basic cosmic truths driving the increasing size of car tailfins in the 1960s.

A more promising effort involves looking for radio signals sent either to contact us or something along the lines of an alien version of South Park that just leaked out. The search for extraterrestrial intelligence, by means of radio transmissions (SETI), began in 1959. Today, the effort continues with multiple programs, equipment dramatically increased in capabilities, and amateur participation: anyone with a computer connected to the internet can contribute. It just seems right that the neighbors would phone up and introduce themselves.

We might not need to find alien civilizations by telescopic or radio searches. Perhaps they have or are visiting us. In the United States, between 2001 and 2020, we had an annual average of over 8,000 unidentified flying objects (UFOs) sightings reported to either the Mutual UFO Network (MUFON) or the National UFO Reporting Center (NUFORC). This location and time period were utilized due to the availability of statistics. Reported sightings have occurred before this, and reports of unidentified arial objects are a world-wide phenomenon.

Various polls taken during UFO presentations, at colleges and convention centers, find that about 1 person in 10 who believes they have observed an unknown arial phenomenon report having done so. This jacks up annual

UFO sightings in the US alone up to 80,000. Additionally, a great many people have not heard of MUFON or NUFORC and would not have any idea where to report their experience and have it counted statistically. I found the apparent number of people who believe they have seen some sort of aerial object they were not able to explain to be personally shocking. I had assumed only an occasional person would think they saw something they would consider to be an unidentified flying object.

US Air Force projects concerned with UFOs reported approximately 1.8 percent of examined cases as not amenable to explanation. Other estimates range upward to 20% with Costa and Costa preferring a 6% estimate. The Air Force would have liked to have explain them all, and the UFO people would have liked for all of them to have been piloted by aliens. Sightings vary from those that are clearly explainable to those that, given available evidence, are clearly not explainable, with no hard boundaries between these categories. If even one percent of sightings are truly unexplainable, that does not indicate that they are alien craft. We are, however, left with a great many events in the US alone that the Air Force and concerned, recognized scientists cannot account for.

Dallas Campbell, a science broadcaster associated with the BBC, presents a collection of reported UFO incidents in "Flying Saucers: A Brief History of Sightings and Conspiracies," found in Jim Al-Khalili's book *Aliens*. In one event, in 1947, Kenneth Arnold was piloting a small private airplane during daylight hours when he saw what appeared to him to be a group of nine flying objects that he could not identify. During the timeframe in which the objects were being observed, Arnold could also see another aircraft, a DC-4, at an estimated distance of 15 miles. This identified aircraft is mentioned to establish that Arnold was familiar with what contemporary aircraft looked like and was not totally out of his element.

Arnold estimated the speed of the unidentified objects, based on distance covered with reference to two mountains of known position, to be somewhere around 1,200 miles per hour, a speed not possible for known aircraft at that time. Human sensory abilities are notoriously questionable. Some sort of input is received by a sense organ and then transferred to the brain for interpretation. It was daylight, so Arnold may have witnessed glinting sunlight in some fashion or another. Or, he may have witnessed alien spacecraft complete with an onboard alien crew. Who knows?

A second incident, discussed by Campbell, took place at an airbase in Britain being used at the time by the US Air Force (USAF). A USAF security patrol reported to have observed a light-emitting object that moved and then disappeared at great speed. Details of the incident include a nearby lighthouse that could have accounted for the light in conjunction with observing personnel imagining movement. Other aspects of the incident, however, make explanations considerably more difficult.

The object reportedly made a return appearance two nights later, with multiple people involved in both sightings. Additionally, in the initial sighting, the object was reported to have been observed on the ground and close enough that one person claims to have actually touched it. Taken in totality, the incident can only be accounted for by either it actually having involved an unknown light-emitting object, or mass hallucinations and hysteria, or the patrol having agreed upon the presentation of a made-up story. Once again, who knows?

If touching an alien spaceship is not bizarre enough, Campbell also includes the case of Betty and Barney Hill, who believe they were abducted by aliens. Chris French, Psychology Department at Goldsmiths, University of London, writes: "It is unclear exactly how many people have claimed to have conscious memories of being abducted by aliens, but the figure is likely to run into many thousands." French believes plausible counter-explanations exist for these beliefs and points out that there are people who "believe that they were Cleopatra or Napoleon in a past life."

Military aircraft have a history of reporting UFOs. Both sides in WW II had reports of glowing objects labeled "foo fighters" by U.S. pilots. Any reported incidents based solely on human sensory input are reasonably questionable due to the limitations of our sense organs and our propensities for self-delusion and hallucination. Reports exist, however, of unidentified aerial objects simultaneously observed by Air Force pilots, videotaped, and recorded on radar. If an object is visually sighted, recorded on videotape, and at the same time tracked on radar with all of them recording identical places and movements, explanations other than a physical presence are hard to come by. If the object performs in a manner beyond the capabilities of known aircraft, or anything a human occupant could survive, explanations are indeed in short supply.

In an incident occurring off the California coast in 2004, two Navy F/A-18F fighter jets from the carrier Nimitz were contacted by an operations officer aboard the Navy cruiser the USS Princeton. For two weeks, the Princeton had been radar tracking unknown aerial objects whose performance was beyond the capabilities of known aircraft. The objects would appear at a high altitude, descend rapidly, ascend at high speed, hover, and disappear from radar all in a manner beyond the capabilities of known aircraft.

The two Navy jets were given the location of an unknown object, based on radar, and they headed in that direction. Upon arrival at the given location, an arial object, with no visible means of propulsion, heat signature, or wings was observed. With the jets flying directly at it, the object first moved toward the approaching Navy jets and then accelerated away in a manner one pilot described as "like nothing I've ever seen." In this incident, the video was recorded that presented a "whitish oval object." In addition to the visual sightings supported by video, the item also registered on radar. With whatever it was observed visually, on radar, and videotaped, it was not a cloud or misidentified star or planet, and since the jet fighters could not catch it, it was not some rich person trying out their new private jet.

Visual sightings and radar readings of unidentified flying objects have occurred repeatedly to U.S. military personnel. During Navy training off the East Coast of the U.S., from the Summer of 2014 to the Spring of 2015, pilots from the aircraft carrier Theodore Roosevelt reported almost daily incidents. An F/A-18 pilot stated, "These things would be out there all day," expressing wonderment at how they could be airborne and maneuverable for such an extended period of time, apparently without refueling. In one incident, an official mishap report was filed when a Navy fighter jet had a near-collision with one of the objects.

Enrico Fermi, with reference to why aliens had not contacted us, asked, "Where is everybody?" If the preceding is any indication, the answer is: All over the fucking place. All of which raises a reasonable question: Why would aliens, if they are here, go to all the trouble necessary to get here? They must have an agenda or agendas inspiring or driving them. A scientific search for knowledge could certainly be one motive. Another inducement, and perhaps of greater significance, could be to study a civilization going through a bottleneck that all developing civilizations must transit. The apparent interest in military activities may reflect a concern for who self-annihilates and who survives.

In addition to our species having entered a bottleneck, there may be any number of behaviors and circumstances of sufficient interest to justify the trip. It may also be that we are not particularly special, but are the subject of study because our home planet is reasonably convenient and any developing civilization is of interest. Or, it could be that we are approaching some major turning point in which we create a generalized artificial intelligence equal to or superior to our own and enter a new civilizational phase, or are approaching a time when civilizations typically leave their biological form behind and become artificial. Or, it may be that they are making an overall study of the planet's flora and fauna, and we are not center stage but merely one member of the cast.

If aliens are here, and they are doing what we would consider to be a scientific study, they would probably be interested in a broad range of topics, the pursuit of which could possibly involve an entity or entities exiting their conveyance and taking a physically involved approach. When Europeans went to the Galapagos Islands, they collected and studied samples. It makes sense that space aliens would engage in comparable activities.

Grinspoon gives consideration to what has been labeled cattle mutilations, referring to cattle that have been found dead with body parts removed with surgical precision: examples of this have been reported in 8 states in the U.S. and in South America and Europe. These incidents have been termed mutilations, but could also arise through the collection of samples.

Grinspoon states that the cows in question are "found with various parts removed with surgical precision – most often the tongue, the brain, other soft tissues in the face and upper body, and the genitals and anus." He further reports, "There is never a trace of a perpetrator, no footprints or tire tracks-..." One incident occurred with snow covering the ground, and yet the perpetrators left no trace of their activities and presence.

He has spoken with persons who have personally seen the deceased animals and have viewed photos taken of them. He states: "I cannot dismiss the impression that something strange, twisted, and hard to understand has actually happened here." With reference to an interview with a rancher and viewing her photos, he writes: "My eyes were unavoidably drawn to the smooth incisions where the poor beast's missing genitals had once been. It was carved up in a way that looked careful and deliberate – and sick."

Grinspoon must be well aware of the value of the selected sampling of subjects being studied. Humans studying and doing research collect samples. He apparently just does not connect the dots when it comes to aliens also behaving in this manner. "I'm not convinced by any of the 'rational explanations.' Yet I see no reason at all to link cattle mutilations with extraterrestrial life." He presents no possible explanation for the incidents; he just knows this is not how space aliens would act. They might more reasonably phone up through the SETI program and not just skulk in like Jane Goodall at Gombe. As one chimp said to another, "I do not believe this Jane Goodall entity actually exists, but I do not have a clue as to where all these bananas come from."

Much of the scientific establishment rejects the notion that aliens have come to this planet. Grinspoon states, "How, then, are we to interpret the fact that, as yet, we have found **no scientifically accepted evidence for the past or present visitation of intelligent aliens** (the bold print is my doing)?" Military aircraft have reported viewing unknown aerial objects since WW II, and we now have simultaneous visual, videotaped, and radar perceptions of the same objects doing things of which no known aircraft are capable. We have precisely dissected cows with no trace of who or how this was accomplished, even with snow on the ground to preserve a record of activities. If aliens are here, they are not putting a great deal of effort into staying concealed, they just are not formally introducing themselves.

Then there is the matter of the thousands of people who believe, or claim to believe, that they have been abducted by aliens. This often includes being transported to an alien ship and subjected to medical study. In many cases, the aliens have an interest in human reproduction to include sexual behavior, with some people claiming to have engaged in sexual activity with them. If aliens are conducting scientific research on this planet, it is reasonable that they would be interested in people, how we reproduce, and the intricacies of human sexual behavior.

Unlike the interactions of military aircraft with unidentifiable aerial objects, there are no videotaped records of these interactions, nor anything comparable to simultaneous radar sightings. Corroborating evidence consists of memories, in some instances of memories being enhanced through hypnosis. Once an idea has been put into our heads, through suggestion or hypnotic procedure, the tendency is to fervently believe in it.

If aliens are rambling around studying stuff on this planet, they could reasonably, as suggested, concern themselves with our species and its behavior, including sexual behavior. It is also true that we are capable of believing all sorts of things and space aliens and people having sex does sound a bit like people considering themselves to have been visited at night by an incubus or succubus and forced to have sex (for the unenlightened, incubi and succubi are sex workers from Hell).

Regardless of whether or not space aliens have ever had sex with people, they could have an interest in how our species does gender. As previously discussed, human males are an evolutionarily altered version of the female, originally of value to their community in the seizing and holding of territory necessary for survival. With nuclear weapons, male aggressiveness goes from being an asset to promoting extinction through nuclear war. Passing through the bottleneck and avoiding nuclear self-annihilation may necessitate the deconstruction of maleness. Just as all civilizations may experience a bottleneck similar to ours, intra-community competition may result in the evolution of something comparable to maleness in our species. How this is dealt with may play a role in whether or not a civilization survives its bottleneck.

If aliens are here, and have gone to all the trouble to get here, why would they not announce their presence and introduce themselves, as opposed to just lurking about and permitting us natives to deny their presence? The answer to that question lies with the effects of European contact on native societies.

European technology, weaponry, and religious assertiveness shattered native beliefs in the worth of their own civilizations. Their sense of specialness and superiority, and belief in their singular and unique relationship to forces beyond the physical world, was devastated. The results were elevated levels of despondency, alcoholism, and suicide, and an uncertainty of their ability to guide their group meaningfully into the future.

Entities from beyond our world could reasonably be knowledgeable concerning the likelihood, and perhaps certainty, of this response from Earth civilizations should they introduce themselves. Visiting aliens, given their ability to traverse interstellar space, would be even more technologically advanced, in comparison to Earth civilizations, than the Europeans were with reference to native societies. Their technology would likely seem incomprehensible (Navy pilots were astounded by the unknown aerial objects' ability to stay up and maneuver all day with no apparent refueling), and they probably would not adhere to the canons of any of the Abrahamic religions. There goes our sense of self-worth, specialness, and superiority.

Foo fighters and cut-up cows can be explained away, or at least ignored, without resort to speculations about space aliens. What is needed, to do the trick, is debris from a wrecked something-or-other with an unmistakable non-Earth origin.

On February 4[th,] 2023, a Chinese spy balloon was shot down and fell into the ocean off the coast of South Carolina. The balloon was big. Thousands of pounds of material were scattered over about a square mile of the ocean bottom. All the wreckage of interest to the United States government was recovered from the site.

With the shooting down of the spy balloon, the North American Aerospace Defense Command (NORAD) became interested in what else was up there. Its radars had been set to detect items above a certain size and faster than a certain speed. They did not want weather balloons or ducks setting the things off.

With the spy balloon incident, radars were adjusted to detect smaller and slower items. Results were impressive. Things about the size of Volkswagen Beetles and traveling at wind speeds began to show up. For three days in a row (10, 11, and 12 February 2023), objects were spotted, rocketed, and thereby downed by fighter aircraft. Pilots of the aircraft doing the rocketing, at around $400,000 a pop, said that they appeared to be metallic and broke into pieces when hit.

Just as with the spy balloon, attempts were made to recover wreckage. Debris from the first item fell off the coast of Alaska on sea ice, material from the second fell in the Canadian Rockies of Yukon Territory (it had been shot down by a Canadian jet), and debris from the third fell into Lake Huron. Nothing was recoverable from any of these three items. It just could not be found. It also appears that no photos were taken of the objects before or debris after rocketing them. All three pilots apparently just forgot to take pictures.

Justification for rocketing the three objects was that they were at an altitude that presented a danger to civilian aircraft. NORAD is apparently not seeing any further VW Beetle-sized items. Have they set their radars back to prior settings? Are objects being seen but ignored? Was the sky simply crowded with objects only when the radar was first adjusted due to the spy balloon incident? If these things were a danger, justifying the use of costly rockets, would other similar items not also be dangerous?

One reason for mentioning the spy balloon is that if whatever was desired from its wreckage could be collected, it is highly unlikely that nothing could be managed for the wreckage of the three additional items. They were smaller, being the size of a small car, with the material from one landing on sea ice, it seems unlikely that nothing could have been found and gathered.

It also seems unlikely that no photos were taken. Rocketing is done from a distance, but aircraft would likely have flown at least somewhere near the items prior to rocketing them. The jets could also have proceeded in the direction of the debris fields, following the rocketing, and taken photos. Pilots described the items as being metallic and breaking into pieces. If they could have been seen well enough to foster that description, onboard cameras could have captured something.

I do not think it likely that we have alien bodies or even mechanical items being worked on to gain knowledge for our own purposes. My suggestion is that we have some material and photos that confuse the crap out of those in-the-know. The easiest thing is to deny everything. This is telling a lie, and once a lie is presented, it is difficult to change anything.

I have, rather unimaginatively, been assuming that the spacefarers would be some alien form of biology. While it is comforting to think of little green folks climbing out of their spaceship, the aliens may have abandoned biology and gone artificial. Something we may be on the verge of doing. Biological creatures from this neighborhood do not do well beyond the protection of our planetary atmosphere and magnetosphere. This is probably true across the board.

The Earth's magnetosphere and atmosphere filter out subatomic particles and electromagnetic radiation from the sun and other stars. Without this protection, these particles and radiation harm DNA and increase cancer risk. Beyond this day-to-day cumulative harm, the occasional solar coronal mass ejection would directly kill unprotected persons. In addition to this, being in the weightlessness of space leads to its own damage.

With the relatively short exposure time of today's astronauts to the vagaries of space, in comparison with the time necessary to travel to Mars, and beyond that, the incomprehensible interstellar distances, they, of superb physical condition, have experienced weakened bones and muscles, impaired vision, nausea, and insomnia. A consequence of no gravity is bodily fluids sloshing around rather than being pulled down into the lower extremities. Bodily fluids spreading into the chest and head, including the inner ear, lead to a lack of balance and nausea, potentially resulting in vomiting. The weightlessness of orbiting in space affects every aspect of the body. Bone and muscle deteriorate. Hearts shrink and cardiac capacity diminishes.

Regardless of our soft, squishy bodies' lack of suitability for exiting Earth's atmosphere and gravity, enterprising humans are planning to make it a possibility for just about anyone with the bucks to be Buck Rogers. Virgin Galactic, apparently with minimal restrictions concerning age and health, books passengers for what amounts to four minutes spent in orbit high enough to be considered space. Orbital Technologies, a Russian company, plans to construct a space hotel for 5-day stays. We are apparently on our way, suitability or not.

Dumping our biology and going artificial opens up the possibility of space travel, and we are moving in the direction of artificial anyway. There is no magic in the human body. It is entirely a matter of chemistry and entirely amenable to science and scientific manipulation. Expanding knowledge is leading to increasing control of disease and degenerative processes. Being able to fix things permits the repair of damage arising from short-term non-Earth travels. Longer adventures require going beyond our current biology and moving strongly in the direction of the artificial.

Regardless of whether or not the human future will entail adventures in space, extension of individual lives, in the here and now, requires replacement of body parts. Worn-out knees, hips, and ankles are currently commonly replaced. Hearts and other internal organs are harvested from the deceased for use as replacement parts. Pigs are being bred with the hopes of harvestable hearts for humans. A mechanical left ventricular assist device is currently being permanently implanted to keep persons with weakened hearts alive and functional.

Growing and manufacturing replacement body parts will become more common, with the results becoming more reliable. The biological body equipment we are born with represents no upper limit of functionality and durability. It should be possible to engineer replacement parts that last longer and function better than those with which we have evolved. Our evolved biological selves may be a short era in the lifespan of our species. It may be common for sociality advanced alien lifeforms, in general, to have had a relatively short period of biological evolution ending with artificiality.

We are approaching the creation of a generalized artificial intelligence surpassing our own. It is a matter of when, not if. Perhaps, along with an improved body less susceptible to the ravages of age and space travel, we can artificially enhance our cognitive capacity and move beyond proto to actual intelligence.

Elon Musk, famously of Tesla, is working on a brain/machine interface that would connect the human brain directly to an artificial intelligence. Musk's thinking is that, with developments in artificial intelligence moving in the direction of eventually surpassing human abilities, we need to either merge with machines or get left behind. This work involves implanting electrodes into the human brain and, in addition to raising the possibility of a human/machine interface, has the potential to alleviate some current physical problems, such as Parkinson's Disease.

Musk's system, known as Neuralink, has had testing with rats and monkeys in which the animals were able to move cursors on screens. Approval has been given by the United States Food and Drug Administration to move on to human volunteers, and this is now occurring, with the hope of alleviating human suffering concerning such conditions as the aforementioned Parkinson's Disease.

Just as Musk's efforts with electric vehicles are not unique, other researchers are also working on brain interfaces. University of Washington researchers have made it possible for people to play games together using just their thoughts. The process does not involve surgical implantation of electrodes, but rather wearable electroencephalography caps that pick up electrical activity in portions of one person's brain and are linked to wands that generate electrical activity in another person's brain.

University of California, at San Francisco, researchers have captured neural signals from people talking and, by computer, produced speech. Given the length of time aliens capable of interstellar travel have had, beyond the age of science on this planet, to work on this, or anything else, what they are capable of will most likely, at least initially, be incomprehensible. We have no idea what would disembark from an on-the-ground flying saucer, or if the saucer itself would be the lifeform.

If the possibility of which has been raised, humans have been taken aboard a non-Earth vessel, an alien entity may have structured something (little green men or whatever), based on brain scans of the humans, that was speculated to possibly be the most propitious way of interacting with them. French reports on a professed abductee claiming to have been induced to have sexual intercourse with an attractive female alien who made dog-like barking noises during the sex. Who knows where that could have come from? A fertile human imagination? Another, admittedly highly speculative possibility, is that alien research efforts confused brain activity concerning sex with a love of and interaction with dogs.

Beyond the possibility of interactions with space aliens, we do not have a clue as to what will become of us. We could just as easily nuke ourselves into extinction, go artificial, or you fill in the blank _____. What is highly unlikely is that things for us will stay the same. What is highly likely is that if we do actually meet some space aliens, they will have agendas, problems, and baggage. My guess is that civilizations, both ours and alien, do not solve their problems so much as trade them in for new ones. If they did not have agendas, they would not have gone to all the trouble of showing up. And, they will arrive with baggage and their own problems.

Enrico Fermi famously asked the downer question, "Where are they?" His position being that if the cosmos is populated, we should know it by now. He then provides an explanation for why, even if they do exist, they are still not here and never will be: the distances involved and the time it would take to traverse them are simply too great given that nothing can exceed the speed of light. From this perspective, maybe they are and maybe they are not out there, but we cannot, for better or worse, get at each other.

Fermi may be being overly pessimistic. The aliens may not be wrapped any tighter than we are and will just go for it, one way or another. A spaceship, adequately shielded from the vagaries of interstellar travel and rotating to provide artificial gravity, could manage the extensive travel time by means of a multigenerational crew. Finding folks who can get along together for even a couple of months is certainly difficult for us, and quite possibly would also be for the aliens. In this case, they would need to get along over a lifetime. And that is just the start; their offspring and possibly their offspring's offspring would need to get along over their lifetimes as well. While the initial crew could be carefully selected, on down the generations everything would be out of control. Good luck with all that.

Another possibility is that of a crew with extended life spans. Once again, considerable shielding would be required, and some means of inducing artificial gravity would also be necessary. As with the previous possibility, a crew would need to cooperate and get along over a really extended period of time, something that at least our species is not all that good at. This starship concept is a Noah's Ark sort of confection that I do not think is much more realistic than the multigenerational concept.

A science fiction standard answer to long travel time is for the crew to be in some sort of hibernation or suspended animation. Shielding adequate for crew protection could then be limited to a relatively small area where the occupants are stored; and there would be no need for artificial gravity as everyone would only wake up, presumably, immediately prior to the journey's end. And, with everyone medically conked out, there is no concern for interpersonal conflicts, or keeping them adequately occupied and amused. If the technology were available, this might be how the aliens and we get around.

Another, and perhaps more likely possibility for both us and any alien travelers, rests with the crew having gone artificial. With no delicate, biologically evolved life on board, the ship would not need so much protection from the vagaries of interstellar space. The crew could possibly stroll around both inside and outside the vessel for scientific research or just to look at the stars. My guess is that if the aliens show up, they will be beyond their biological origins, and if we show up there, we will be too.

Another concern centers on the extent to which alien crew members would maintain an individual status. So far, on this planet, only hardwired varieties of life have attained the supraorganic, with individual entities merging into a new level of life. Our softwired selves are strongly in the direction of the supraorganic, but have not taken a final big step of losing our individuality to the group, as some species of ants have. Whether or not we will ever take the final step and go Borg or Hive Mind is totally unknown. We should, however, be prepared to meet aliens that are not only artificial but also supraorganic. Under these circumstances, the entire crew, and possibly the ship itself, would have merged into a single entity.

Should all or part of our species choose to go artificial, perhaps as a means to extend lifespan and/or gain suitability for space travel, managing the transition to artificiality of the human brain, with its multilayered, evolved complexity, would present the greatest challenge. Going back to the "engineer-designing-a-car" scenario, loading our brain into something artificial presents the opportunity to chuck stuff out.

As our brain evolved and behavior became more complex, new layers were added on top of old, with the old remaining in place and continuing to function. Our brain stem remains from reptilian times and continues to regulate bodily functions such as respiration and heart rhythm. As an example of how brain areas remain fundamentally the same, the brain stems of chimpanzees and humans are difficult to distinguish.

The most recently evolved area is the frontal lobe of the cerebrum, which is associated with thinking, planning, and judgment. Additionally, between the brain stem and the frontal lobe, there are all the intermediate, post-reptile, and prehuman layers. Presumably, if we went artificial, the brain stem could be discarded and replaced by some mechanical control mechanism or mechanisms. Even though we may only have a precursor of intelligence, we would certainly want to keep what we have and transfer the frontal areas of the brain. A great deal of uncertainty exists for intermediate areas.

As the ancestors of our species evolved, new behavioral attributes were commensurate with increasingly complex and layered brain structure. We fancy ourselves as sapient and wise with our behavior under the control of the newly evolved areas. In reality, the old parts continue to chug along underneath, providing drives and sponsoring behavior with the new parts, which we prize as making us rational, often serving only to justify doing what the older parts want.

The older brain areas, while sponsoring possibly undesirable behavior, may also provide us with a sense of meaning and a desire to stay alive. The Spock Fallacy, originating with the Star Trek character, presents the notion that if we could just be totally logical and free ourselves from emotional irrationality, we could lead untroubled and enlightened lives. The fallacy is that without drives originating in older brain areas, nothing has meaning. In and of itself, it is not rational to keep breathing or eat dinner. Our underlying irrationality might lead us to fuck up our marriages, but without bonding propensities from our mammalian past, we would not get married in the first place.

If you load chimpanzees into a spaceship and send them to another planet, they will still be chimpanzees when they unload at their destination. If you load all the stuff in the human brain into a machine, you end up with a machine that behaves like our species of ape. How do you pick and choose brain material so as to end up with something at least somewhat sensible and yet not excessively prone to suicide? It is a conundrum. What should our future be? We are going to manufacture it.

This chapter has dealt with the cosmos' commonalities. Chapter Nine will deal with the trials and tribulations of trying to grasp the cosmos' complexities, of which we are a part.

CHAPTER 9

Crap
Space Is Not Empty

Space is not a perfect vacuum. Space is loaded with all sorts of crap from forces and fields, and subatomic particles, to massive black holes and sheets of galaxies. And, all the crap is not in space. A very adequate load of it lies within academia. If you look at the history of any academic pursuit, social science, or hard science, you will find that changes in ideas and beliefs, big and small, have occurred over time. If you choose a particular historic period (any time in the past will do), the leading people in every discipline always think they are close to knowing just about everything to do with their subject. There might be conflicting ideas, but the thinking would have been that this would be settled, and then just about everything will be nailed down.

These beliefs pervade the sciences and, viewed from an historical perspective, are always wrong. If core beliefs are not replaced in their entirety, they will, at a minimum, be renovated in response to new data. Going from 1700 to 1800 to 1900 to 2000, changes have occurred across the board in all sciences. When we go from 2000 to 2100, it will be the same story. Some disciplines will experience the collapse and restructuring of basic beliefs, and all will at least be tuned up. An enormity of big shocks, along with endless updates, remain to be lived through.

Associated with the belief that we are on the edge of grasping just about everything is the concept that the only thing needed to achieve total closure is an incremental increase in knowledge. I have never heard of a scientist or professor stating that the paradigm on which she or he has built a career and taught in the classroom is fundamentally flawed and in need of replacement. Yet, this is the reality in which many academics are embedded.

In addition to accumulating incremental change, disciplines, in fits and starts, experience revolutionary change in which their worldview disintegrates in the face of accumulating discordant data and is replaced by another. I have been privileged, as previously mentioned, to witness this event in geology, am under the impression that it is currently taking place in sociology, and intend to argue that cosmology will eventually undergo this process.

With geology, the change to the current continental drift paradigm took decades to work through the discipline, with the old school devout adrift with incomprehension. As the saying goes, "Science progresses one death at a time."

As previously discussed, sociology is currently riding the struggle bus of paradigm change. While psychology and anthropology have incorporated biology into their worldview, sociology continues to fail to do so, as evidenced by the following statement from Giddens and his associates: "It has been difficult to find compelling, substantive data on the interaction between biology and experience." With biological variables falling outside the amalgam of the sociological standard model, "compelling, substantive" evidence for biology influencing human behavior remains either invisible or, if as apparent as the accumulating geological data that challenged the fixed continents paradigm, incomprehensible to the devout.

Further down, Giddens et al. reveal the confusion in contemporary sociology. "Gender differences are not biologically determined; they are culturally produced." In other words, with reference to the sociological standard model, the behavioral differences between females and males are entirely a matter of socialization. Soon after this, we learn that the sociological paradigm, as happened with the fixed continents paradigm in geology, is evidencing internal fracturing with the accumulation of new data. "Despite critiques of the 'nature' perspective, the hypothesis that biological factors determine behavior patterns cannot be wholly dismissed."

Confusion over the role of genetics in human behavior is found throughout the Giddens et al. introductory sociology text. We learn, for example, "sexual orientation results from a complex interplay of biological and social factors, not yet fully understood." Biology, however, quickly vanishes: "Like all behavior, heterosexual behavior is socially learned within a particular culture."

A difficulty with adding genetic factors to a mix with sociological factors lies with sociology's history of presenting human behavior as entirely the product of culture and the behavior of all other animals as "instinctual responses to the environment." With the entrenched idea that behavior was either the product of culture or the product of biology, the necessity of understanding how a mix of the two works is difficult: the mix of biological and social factors involved in sexual orientation is "not yet fully understood."

Geology has very clearly experienced a scientific revolution. Sociology is currently in the process of a paradigm change. My prediction that cosmology will also experience this process is based on the enormity of new incoming information. Instruments are peering farther into the cosmos, with data coming in from one end of the electromagnetic spectrum to the other. These circumstances are optimal for a shake-up of core beliefs.

Common practice is to exert considerable effort toward the preservation of a discipline's current standard model in the face of incoming conflicting data. Complex systems of cycles and epicycles kept the Earth-centric model of the cosmos alive for a bit longer than might have been appropriate. Guth's inflation concept, in which a massive expansion occurred in a sliver of a second immediately following the Big Bang, explained how it was that the universe did not quickly collapse back on itself, thereby wrecking the Big Bang. Most recently, the James Webb Space Telescope found items at such a distance that they should not have had time to form. The response has been to put effort into understanding how these things could have formed more quickly than we had thought possible, rather than following their clear implications and junking the current Standard Model.

An additional factor suggesting the likelihood of paradigmatic change in cosmology lies with the paucity of hard data available for theory construction in the not-so-distant past. The above-referenced massive influx of data lands on a frail foundation short on facts and long on creativity and imagination. A lack of knowledge does not interfere with or repress our species sapient view of itself or its fabulous creativity. It should be kept in mind that the species of ape providing the contemporary Cosmological Standard Model is the same creature, using the same cognitive organ, that gave us Catholicism.

Contemporarily, Big Bang Theory is central to cosmology's standard model. Big Bang holds that everything emerged from either a singularity, a primeval atom, or possibly a wad of stuff the size of a grapefruit or your personally favorite citrus, or even nothingness that happened to be in an unstable state. Once coming into existence, an ongoing expansion began and continues to this day in which EVERYTHING, including not only all the matter in the cosmos but space and time themselves, achieves actuality. It did not expand into anything because nothing existed prior to this incident for it to expand into. To borrow from Georges Lemaitre, "it was a day with no yesterday."

If, with the Big Bang, the universe emanates from something or other, the beginning has, in reality, been put off to the origination of whatever it was that it sprang from. A. S. Eddington postulated a primitive atom "having an atomic weight equal to the entire mass of the Universe." Under these circumstances, the Universe already existed, but was packaged differently, and we remain clueless as to its origin. Lawrence M. Krauss solves this conundrum by assuring us that we do not need anything in order to get something.

Krauss points out, "The theories that underlie much of modern physics – all suggest that getting something from nothing is not a problem." He further illuminates our understanding of these matters by pointing out that "The structures we can see, the stars and galaxies, were all created by quantum fluctuations from nothing." Krauss contends that "nothing" can be unstable and, under the sway of quantum fluctuations, may burst forth into a universe. He further contends that it is actually essential for the universe to have come from nothing.

Regardless of the complexities of origin, most cosmologists since Edwin Hubble's work in the 1920s and 30s have accepted the Big Bang as real. Krauss assures us that "all evidence now overwhelmingly supports" this doctrine. Fractures, however, may be forming. John Boslough, in his work *Masters of Time: Cosmology at the End of Innocence*, reports that "almost everyone," including himself, had accepted the reality of the Big Bang Theory, but continues on to state that by the mid-1980s, it was becoming "apparent that something was wrong."

Hubble laid the foundation for the Big Bang back in the 1920s. Utilizing the new Hooker Telescope on Mount Wilson, he revealed that the fuzzy splotches known as nebulae were located not within but beyond our galaxy. At that time, the entire universe was generally thought to consist solely of our galaxy – the Milky Way. With his discovery, we came to realize the Universe was much bigger than we had imagined.

Hubble found evidence of the vast distances separating us from what we were beginning to recognize as other galaxies. He accomplished this by means of a type of variable star known as a Cepheid that the Hooker telescope had the capability of finding in the newly discovered other galaxies. This category of stars can serve as distance markers because their brightness is related to the length of their variability cycle. By checking the variable time period, we know how bright the star would be in its own neighborhood. Brightness declines inversely with distance. The extent of decline in brightness from the actual brightness then indicates distance.

As a complicating factor, if the object emitting the light is moving away, the light stretches out or redshifts to an extent based on the speed of departure. Hubble found a relationship between distance and the extent to which light was redshifted. The farther the object, the greater the redshift. Based on this, the cosmological community concluded that the Universe was expanding and that this expansion led to everything moving away from everything else. It was like raisins all moving away from each other in a loaf of bread expanding as it baked.

With moving away, the speeds added up, making the farther objects move away from us faster than the nearer objects. That was why redshift was increasing with distance. The idea that everything was moving away from everything else implied that it must have all started somewhere with everything gooped together. Thus, Big Bang Theory coalesced.

It is not, however, possible to find the location of the original item because nothing existed prior to its emergence. This means that it appeared everywhere and nowhere. The Universe was not expanding into its surroundings. There were no surroundings to expand into because the Big Bang was creating everything, once again, as previously stated, to include space and time. The recessing items were not moving away through space. Space itself was expanding and carrying the items with it into a nowhere that did not exist.

Even though the location of the Big Bang was unknowable, Hubble attempted to establish the speed at which items were recessing and use this estimate to compute the age of the Universe. The initial results Hubble came up with were unrealistically low. They placed the age of the Universe at less than 2 billion years when geology was establishing the age of the Earth at around 4 billion years. One factor producing the obviously wrong estimate was an error in the distances to items beyond our galaxy. Greater and more accurate distances produced a succession of older ages, with widespread general agreement remaining elusive.

The difficulty of computations is increased by "peculiar motion." In addition to recessional velocity, a great deal of gravitationally sponsored motion exists. The mass of two physically close galaxies generates a gravitational attraction sufficient to result in those galaxies moving toward each other faster than space is expanding. Andromeda, for example, is moving toward the Milky Way faster than expanding space is moving them apart. This "peculiar motion" is not only evidenced by locally grouped galaxies. The gravitational attraction of enormous galactic clusters results in widespread motion over large areas as well as the more local occurrences.

If peculiar motion is sorted out of recessional velocity, it should be possible to run everything backwards and compute how long ago the Big Bang occurred. This has proven to be a bigger problem than might be supposed, as different approaches produce different answers. One method utilizes Cepheid stars, another the merger of two neutron stars, still another makes use of supernovae, and a fourth involves an analysis of the cosmic microwave background radiation. The important fact is that every method produces a different result.

Initial estimates were hampered by difficulties in measuring distances to the newly recognized galaxies. Back in 1931, Hubble's first paper on the subject placed the age of the Universe at 1.8 billion years. With more accurate measurements, using RR Lyra stars rather than Cepheids, distances and age were doubled. With the greater capabilities of the new two-hundred-inch telescope at Palomar, Alan Sandage, in the 1950s, pushed the age of everything to 5.5 billion years. In 1975, Sandage, once again, jumped the age of the Universe to "approximately 15 billion years." Further estimates, by an assortment of astronomers and cosmologists, varied from 10 to 20 billion years. Guesses were all over the place.

A difficulty facing those attempting to determine the age of everything was learning what was happening to the speed at which the expansion was occurring. It was widely thought that the expansion rate would be slowed by gravity. Through the study of supernovae, Saul Perlmutter, astrophysicist at the University of California, Berkeley; Brian Schmidt, astrophysicist at Australian University; and Joshua Frieman, astrophysicist at the University of Chicago, came to the opinion that the expansion velocity was actually increasing. The explanation: a mysterious and unknown force labeled "dark energy" was counteracting gravity and driving the increasing rate of expansion.

To compute the age of the Cosmos, it would now be necessary to take into account the initial expansion rate when all the material was closer together and gravity would have been more powerful, as well as the current rate when gravity would have weakened with growing distance and the mystery force accelerating expansion would possibly have gained in power. As if this were not problematic enough, another issue of enormous complexity has arisen.

To keep the Big Bang Theory afloat, it became necessary to postulate, as previously mentioned, a high-speed inflationary period more or less instantly following the Big Bang. Back in 1980, Alan Guth of MIT proposed this inflationary period, and many cosmologists jumped on board. Among those accepting the concept, total agreement existed that it was not a matter of material moving through space, but rather space itself expanding and carrying the material with it. Interestingly, this postulated inflation occurred well in excess of the speed of light. While the speed of light is generally considered to be the speed limit for the Cosmos, apparently, expanding space has been granted an exemption – possibly by God.

As a further complicating factor, space has grown to the point at which there is now so much of this expanding stuff between the cosmic raisins that, according to Australian cosmologists Charles H. Lineweaver and Tamara Davis, circumstances now exist in which some of these raisins have attained a recessional velocity exceeding the speed of light. This is not the only time recessional velocities beyond the speed of light have been heard of. John North also mentions recessional velocities exceeding the speed of light in discussing Stephen Hawking's theories. So, in an initial inflationary period, material moved faster than photons, and we currently have material recessing faster than light due to the quantity of space that has come into existence. Once again, it is not material moving through space faster than light, but rather space expanding and hauling material with it. The emergence of space is apparently something of a wild card a cosmologist can play in support of her or his theoretical constructions.

Krauss backs up Lineweaver and Davis in their notions concerning the natural laws of space growth. He points out that while no material can travel through space faster than the speed of light, "expanding space has no such limits." The objects are, in reality, at rest in space, with that space expanding and carrying them along for the ride. Keep in mind that prior to the expansion of space, from its beginning with the Big Bang, nothing existed for space to expand into. Everything was being created, including space itself, while space expanded.

Areas of space can be empty with reference to material that is substantive and potentially visible, but all space is apparently always jam-packed with fields or energies of some sort or another. Krauss asserts that "an invisible background field, specifically a Higgs Field, permeates all of space." Thomas S. Kuhn has stated with reference to space, "some of its new properties are not unlike those once attributed to the ether."

Scientific concerns for space lay with its properties, such as how light and gravity could propagate through it. Space could not simply be devoid in totality. Some proprietary substance appeared to be necessary. In the 17th Century, Robert Hooke postulated "an all-pervading ether." In the 19th Century, a widely held scientific belief was that light propagated through "ether-like waves of sound through air." Something had to be there as "T(t)here could be no waves on the sea without an ocean of water to carry them."

Albert Einstein launched a new and contemporary view of space in a speech given in 1920, ascribing attributes anointing it as a new version of ether. "To deny the ether is ultimately to assume that empty space has no physical properties whatever." He went on to state that "his gravitational field theory implied that empty space had physical qualities." Einstein argued that there is no such thing as empty space, by which he meant space without a field.

In the 19th Century, physicist Albert A. Michelson experimentally provided substantive evidence that ether did not exist. What did he know? It seems that, when needed, a modern version of ether, known simply as "space," or "field," or nothingness impacted by "quantum fluctuations," could be called to duty. Max Jammer, in concluding his book on the history of space in physics, stated, "The problem of space will have to be classed as unfinished business."

Getting back to determining the age of the Cosmos: if we are going to run everything in reverse and go back to the original item or unstable emptiness from which the Cosmos emerged, we must now take into account the initial inflationary period, and an expanding universe that does so at unknown and varying speeds. Regardless of the difficulties, Charles Lineweaver was certain he could do the math and compute when the Big Bang did its thing and all space, time, and matter began the process of initiating from nothingness or something the size of a _____ (fill in the blank with the name of your favorite citrus). The answer: 13.4 billion years ago. Being modest, Lineweaver, unlike Bishop James Ussher, did not presume to know the day of the week.

Do not stop breathing, but Lawrence M. Krauss states that Lineweaver was off by 320 million years. The real, and even more precise answer is 13.72 billion years. In an "afterward" to Krauss's book, *A Universe from Nothing*, Richard Dawkins refers to Krauss's stated age of the Universe as something that can be accepted "with confidence" and that it is characterized by "stupefying accuracy."

Krauss is not the first researcher to identify so precisely when everything got started. The previously mentioned Ussher, through the diligent research of an extensive body of scripture and religious writings, learned that God had created and cranked up everything on Saturday, October 22, in 4004 B.C. Ussher was not the only theologian attempting to determine when everything began. A problem was that the religious folks, as with the cosmologists, all arrived at different results.

Krauss simplified his search for the beginning of everything by avoiding measurements of distances and velocities and analyzing the precise measurements of the cosmic microwave background that had become available with the deployment of the Wilkinson Microwave Anisotropy Probe (WMAP). As a sociologist, I do not have a clue as to how analyzing the cosmic microwave background can provide the age of everything as precisely as Krauss claims it does. It seems to me on a par with reading chicken guts.

George Efstathiou, director of the Kavli Institute for Cosmology at the University of Cambridge, in disregard of Krauss's stupefyingly accurate SWAG (Scientific Wild Ass Guess), but also based on studies of the cosmic microwave background, estimated the age of everything at 13.75 billion years. After more consideration, he amended this to 13.81 billion years. Unfortunately, cosmologists using exploding stars to make age estimates get different answers than those using the cosmic microwave background.

Anna Ljjas and Paul J. Steinhardt, both associated with The Princeton Center for Theoretical Science, and Abraham Loeb, of the Astronomy Department at Harvard University, assert that the structure of the cosmic microwave background can be interpreted in multiple ways: "it can be made to predict almost any outcome." Who knows?

Dark energy is theorized to be driving an accelerating expansion of the cosmos whereas if only gravity were involved, inflation should be slowing down. In addition to dark energy, we have dark matter. With the strength of gravity that would arise from all the matter that we can detect, a great deal of rotating material, given its velocity, would establish different trajectories. We know galaxies are not flying apart, so there must be material out there that we are unable to detect.

If our species, given its magnificent brain, brilliant theorizations, and technological advances, is unable to get a grip on the material that is out there, that material cannot possibly be normal stuff. That material must be beyond the normal and must reasonably be labeled "dark matter." It has got to really be an unknown type of crap. It is certainly not a regular matter.

Unfortunately, there are hints that an enormous amount of regular crap exists out there that we are unable to detect. There may, for example, be a planet "probably six times more massive than Earth" loitering about in our own Kuiper Belt. If we have difficulty confirming whether or not a planet-sized object exists in the outer area of our own solar system, should we not expect to have difficulty finding material of various sizes at considerably greater distances?

We are gaining in our ability to find more cosmic crap. Brown dwarfs are objects not as big as stars and therefore not lighting up due to the fusion of hydrogen into helium. They are, however, larger than planets: big enough to emit thermal radiation when formed and then slowly cool. In the past two or three decades, it has become possible to detect brown dwarfs by means such as the Wide-field Infrared Survey Explorer Telescope launched in 2009. With around 3,000 now known brown dwarfs, estimates suggest that "the Milky Way contains between 25 billion and 100 billion brown dwarfs."

We are struggling to find Earth-size planets. What about things the size of an asteroid, a walnut, the head of a pin, and atomic fragments? Given our current instruments and theories, we simply cannot identify and locate the normal matter that accounts for how the stuff we can see holds together while orbiting. The fact that it does hold together is proof that the material exists. It is the height of overweening conceit to assume that, under contemporary circumstances, a new kind of matter must exist.

Recall that we do science, religion, and politics with the same brain that does paradigmatic thinking and not rational thought. In every scientific discipline, we think we have the big stuff nailed down and are close to closing in on the rest. In a now-we-know-almost-everything statement, Saul Perlmutter declared, "It almost feels like we're taking our first baby steps as a civilization, toward actually having a model of the universe that will hold up over the next 500,000 years." Good luck.

Stephen Hawking, in a conversation with John Boslough, stated "that by the end of the century he and other cosmologists would have created a single theoretical statement that would describe not only all the physical laws of the universe, but also its initial conditions" (that century has ended). And, further on in the conversation, "My goal is nothing less than a complete understanding of the universe."

Krauss, in *A Universe from Nothing*, states, "(T)he phenomenal progress we have made in the past century has brought us to the cusp, as scientists, of operationally addressing the deepest questions that have existed since we humans took our first tentative steps to understanding who we are and where we came from." Dan Coe, astronomer at the Space Telescope Science Institute states, "We stand on the verge of writing a nearly complete cosmic history." Are we really on the verge of knowing just about everything, or given all that there is to know, are we on the verge of knowing a bit more than nothing? Perhaps it is more like standing at night in the Amazon rain forest and shining a small flashlight around.

Our knowledge of the Cosmos accumulates as our instruments become more powerful and gain capabilities. As previously mentioned, Edwin Hubble, utilizing the new Hooker Telescope, demonstrated that our galaxy was one of many, and our estimates of the size of the Universe incurred a dramatic expansion. Following this, Walter Baade, using the newer and larger Hale Telescope, again hugely jumped our perceived size and age of everything.

There are hints that the preceding trend has not run out of steam. Priyamvada Natarajan, a theoretical astrophysicist at Yale University, in 2018, argued that we are seeing "supermassive black holes, at such distances and so far back in time that, given our notions of the size and age of everything, there does not seem to have been time for them to have formed." Perhaps, once again, things are bigger and older than we had imagined.

The James Webb Space Telescope has been launched and is seeing farther and more precisely than anything preceding it. As Natarajan pointed out and the James Webb Space Telescope demonstrates, if current theory represents reality, items are being found at such distances that they would not have had time to form. The potential is here for a Kuhnian-style scientific revolution.

Big Bang Theory rests on the red-shifting of light in response to distance and recessional velocities. Further consideration should be given to other possible explanations for redshift. Space is not devoid of everything. Einstein argued that a field "could exist in 'empty space' in the absence of ponderable matter." And, of course, a wide assortment of actual matter does exist in space, from stray subatomic particles to giant, high-gravity black holes. Photons may lose energy and their waves stretch out due to the effects of fields and forces as well as whacking into actual crap and the drag from its gravity.

Our instruments, with their current capabilities, peer to the same distance regardless of the direction in which they are pointed. And, regardless of the direction, we are supposedly seeing the edge of our expanding cosmos. If we are seeing just about everything that there is to see, at about the same distance in all directions, this positions us in the center of the Cosmos.

Historically, center stage is where our species normally positions itself. Even though it is highly unlikely that our planet sits at the center of everything, cosmological explanations exist as to how this illusion could be created. We may, for example, only appear to be in the middle because we are doing the equivalent of looking down the edge of an expanding balloon, which just happens to exist in all directions. Very creative and brilliant people can come up with explanations for just about anything.

Perhaps, rather than looking for a beginning, and possibly an end, we should embrace the notion that, for all practical purposes, the Universe is infinite and eternal. That may be the best we can manage given the brain we have to work with. From this perspective, we would look for ongoing cycles.

General agreement places considerable mass, in the form of black holes, at galactic centers. Back in the 1940s, Carl K. Seyfert "noted that the spectra of the nuclei of his galaxies seemed to suggest emission lines of hot ionized gas streaming out at velocities of thousands of kilometers per second." It has been confirmed that "powerful jets" traveling "near the speed of light for up to thousands of light years" emanate from the polar regions of the material at galactic centers. A widely accepted explanation for this phenomenon is "that magnetic fields near the event horizon of supermassive black holes power these jets." The material in question does not, given current theory, emanate from the black hole itself.

As an alternate explanation, perhaps the jets actually originate from the object's interior. The crushing force of gravity created by the mass of material and vortex forces originating with incoming material may crush interior matter into some level of subatomic particles and squirt this out of the polar areas of the item. Rather than everything eventually becoming trapped in black holes, or possibly expanding outward into a forever diluting soup, material squirting out of the polar regions would provide a basis for cycles.

Evidence exists for the accumulation of material to more massive extents than that found central to our own galaxy. Even with substance exiting polar regions, material could continue to aggregate and upon reaching some critical point, disperse in what could be considered a mini-bang. This dispersed substance is then available, along with the black hole ejected material, for further cyclical participation.

John North reports on "massively energetic star-like" sources of energy and particulate matter, with "each pouring out around a hundred times more energy than an entire typical galaxy." It may be that material can come together in amounts sufficient to produce a mini-bang. Rather than ever having had one big bang, we may have eternal and infinite mini-bangs spread across the cosmos like lightning bugs on a summer evening: this is my Lightning Bug Theory of the Cosmos.

The mini-bang perspective, rather than a single big bang, solves the uniformity problem. If there was a single bang, it becomes necessary to explain how it is that the cosmos is so lumpy and bumpy. Noam I. Libeskind, of the Libniz Institute for Astrophysics in Potsdam, Germany, and R. Brent

Tully, astronomer at the University of Hawaii's Institute for Astronomy, explicate the large-scale structures of the Cosmos: "galaxies themselves gather into clusters, and galactic clusters group into subclusters. These galactic superclusters are the building blocks of great filaments, sheets, and voids that constitute the largest measurable structures in the universe."

Valerie de Lapparent, astrophysicist at the Institut d'Astrophysique de Paris, France, in agreement with Libeskind and Tully, found "enormous structures that looked like gigantic bubbles, each some 150 million light years wide," The walls of the bubbles were formed by clusters and strings of dozens and hundreds of galaxies and the interiors were empty. These voids were all over the place. Zooming out, regardless of the perspective, does not achieve the expected homogeneity. Cosmic structure demands multiple bangs and not just a single big one. We may not be closing in on knowing just about everything.

HUMAN NATURE and the COSMOS

PART III
The Unknown

CHAPTER 10

People Land Our Future

Given the groundwork laid down in prior chapters, we now turn to consideration of future possibilities for the planet and our species. My contention is that we are in an era of dramatic and extensive change to include social, technological, and environmental circumstances. It appears that when it comes to our view of human nature, we will continue to drift away from godly and in the direction of apish. Previously cited research indicates a decline in both church membership and levels of religiosity, with these trends likely to continue.

The Abrahamic religions present a scenario in which a god creates everything to include people. Evolution provides a totally different picture. The scientific presentation is eroding religion. If there are problems with the accuracy or factualness of religious beliefs in one area, it opens up the possibility of problems in other areas. These beliefs are presented as the "Word of God." If Genesis is just a story, what about everything else in the Bible?

Females are presented as second-class citizens throughout both the Old and New Testaments. In the Old Testament, God pointed out to Eve, in Genesis 3:16, that her husband "shall rule over you." That precept has yet to be rescinded by Rabbi, Priest, Pope, or Imam. It continues in effect to this very day for contemporary women.

Seeing women as substandard to men does not change with the addition of the New Testament. In Corinthians 7:4, we learn that "the wife does not have authority over her own body, but the husband does." An ancient patriarchy made this stuff up and had it put out by their invented god. As women assume greater power in societies, it seems reasonable that ideas serving a male-dominated ancient civilization will erode.

In addition to an ancient society's denigration of women, Leviticus 20:13 explains that a male having sex with another male is committing an "abomination" and God thinks it appropriate for him "to be put to death," Once again, this news from an ancient patriarchal civilization remains in place for the highly religious to point to and claim as factual. Increasing acceptance of non-traditional gender categories will facilitate the erosion not only of these ideas but of religiosity in general.

My contention is that religion and science present two conflicting explanations of reality. Currently some of the religious are expressing agreement with this by arguing that the young should not be sent to college because it damages them. The problem they have with education is that it introduces science into people's lives in competition with religious ideas. Also, with collapsing male hegemony, some males are turning to religion as it supports male dominance. These efforts will result in a bump in the road but will not seriously impair religious collapse.

As societies change and the scientific perspective expands, the Abrahamic religions, along with religiosity in general, will continue to erode, and church attendance will continue to decline. Abrahamic religious beliefs present ideas commensurate with the needs of an ancient male-dominated civilization. These beliefs oppose female societal participation. Changes occurring in society will continue and will continue to shove religious beliefs out of the way.

In addition to the melting away of religion, climate change will continue to roll on. We have had 10 or 12 thousand years of a relatively mild and stable climate. We do not know how long this would have lasted without industrialization. With industrialization, it is finished off for certain. We are, and will continue to do so for the foreseeable future, zigzag in the direction of a warmer and considerably less uniform climate: storms will be stormier, drought and rain more severe, temperature records set, what had been widely spaced weather events will occur with greater frequency, polar ice caps will continue to melt, coastal cities will experience flooding, and on and on.

The use of non-polluting, renewable sources of energy is increasing. In the United States, renewable use is rising around 3.1 percent per year, with that level of increase projected into the future. Sweden, Norway, and Denmark are leading in the transition from fossil to renewable energy sources. Germany is also strongly heading in this direction. Unfortunately, taking the world as a whole, fossil fuel use is also increasing.

In 2023, the United States' crude oil extraction and exports reached record highs. Our renewable use is increasing, but we are selling a great deal of oil to others who burn it. Worldwide demand for oil is rising. World wide use of coal remains steady due to increasing consumption by India and China. As you may recall, India is attempting to elevate large portions of its population out of poverty by means of energy derived from coal.

The United States is producing a great deal of natural gas. This burns much cleaner than coal and is replacing a great deal of coal in the United States and parts of Europe. Natural gas is, however, composed primarily of methane and loses some of its advantage over coal because of the quantities that escape to the atmosphere during extraction and shipping.

We are in a situation in which the worldwide total energy usage is going up. Energy from solar and wind is increasing, but so is fossil fuel use. The increasing energy extracted from renewables does not subtract from the pollution arising from fossil fuels. Fossil fuels continue to inject pollutants regardless of the production of less harmful energy sources.

Demand for energy rises with both increasing population and elevating living standards. The human population is projected to increase until around the end of the century. This leveling off will occur slowly, and a following population decrease will also occur slowly. There will not be much left of the planet, other than People Land, for a very long time.

Getting the world's already existing population, along with the upcoming additional people, ensconced in some reasonably acceptable lifestyle will require a dramatic increase in energy usage. More than 770 million people on the planet, as this is being written, do not have access to electricity. Of those that do, almost none have the level of luxury found in the United States.

In addition to the increasing number of people adding to energy expenditure, extending the lifetime per person also does. Under these circumstances, rather than the population leveling off and then declining in an expected time frame, the process slows with the percentages of age groups changing: fewer young and middle-aged with more old people.

Medical advances, at their current rate, may be expected to very slowly extend human lifespan. Artificial intelligence raises the possibility of speeding things up. Artificially intelligent devices can look for patterns and connections in enormous piles of data much more efficiently than humans. They can then further generate hypotheses that would have had a possibility of being overlooked by humans. With artificially intelligent devices available to assist in medical research, life-extending advances may be expected to occur more quickly.

In short, it appears that we will be using considerable quantities of fossil fuels well into the future, with their associated climatic damage. Further, if we were to put a dramatic and successful effort into switching from fossil fuels to renewable energy, our environmental problems would not simply rapidly disappear. Climate flows in a geological time frame with considerable momentum arising from past circumstances. The weight of past and ongoing circumstances dictates extensive future climatic unpleasantness.

As if the preceding, slow-moving, population and environmental circumstances are not challenging enough, the prospect of nuclear war raises the possibility of immediate and total environmental collapse, and the creation of genetic damage and birth defects in a world of extensive and long-lasting radioactivity. If it is a decent-sized war, a nuclear winter and mass starvation would be entirely possible. Civilization, as we think of it, could cease to exist. Our species could cease to exist.

If our species were rational, as we typically suppose it to be, there would be no need to concern ourselves with the possibility of nuclear war. The idea of engaging in behavior that would slaughter large numbers of people, possibly destroy civilization as we know it, and possibly even lead to the annihilation of our species, would simply be out of the question. Unfortunately, we are not rational, and old brain parts remain active under the newer areas that we delude ourselves into thinking are in control.

Nuclear war is a straight-line extension of past behaviors. The communities of our chimpanzee cousins raid and kill each other. Ancient human communities raided and killed each other. The ongoing Russia/Ukraine war is a direct extension of this behavior. It is not unthinkable that circumstances may occur under which Russia would employ nuclear weapons. Should those circumstances come to be, other involved nuclear powers may join in.

Russian forces currently (as of this writing) occupy parts of Ukraine. The United States and Western European nations are supplying Ukraine with weapons, typically, in the past, with the provision that they only be deployed against Russian targets in Ukraine, and not against targets in Russia itself.

Ukraine argues that it would be appropriate for them to deploy everything they have, wherever it would be most effective, including inside Russia. Those supplying weapons to Ukraine are shifting toward agreement with this perspective. Russia states that if its homeland is attacked, with weapons supplied to Ukraine by outside governments, it may begin to employ nuclear weapons. Should this occur, we would then be next door to a nuclear World War III. We are actually there already, as in addition to the uncertainties associated with the Russian/Ukraine war, we have the ongoing, unstable relationship between India and Pakistan, along with the Chinese/US interactions in the Pacific.

Masculinity drives war. As previously discussed, maleness evolved to seize and hold territory necessary for community survival. Male violence has been evolutionarily selected for because, in the past, it served group needs. This once valuable product of evolution, under altered circumstances, now threatens human oblivion. War, fought with whatever is available, is standard. Today, that includes nuclear weapons.

A perspective found in both common culture and anthropological circles holds that men evolved as hunters. This positions men as those who fed the community while women hung out around camp and raised kids. As has already been dealt with, women in foraging societies typically provided much of the community's food, with men providing the territory in which it could be gathered. This territory is being secured by means of male violent interactions with neighbors. The male past lies not with hunting, but with war. Our concern lies with the male future.

Human maleness is a construction involving socialization and genetics. With reference to socialization, societies accept and transmit various levels of violence. A struggle is occurring in the United States to reduce what is acceptable. Domestic violence has been criminalized. Over the last several generations, in the United States, violence levels associated with masculinity have fallen.

While the more peaceful society that has been achieved is certainly wonderful, the critical question concerns whether or not this level of softened masculinity is sufficient to eliminate the possibility of a future war in which nuclear weapons are employed. Unfortunately, changes vary widely across cultures, and even where the changes are greatest, the answer appears to be no. My guess is that masculinity will lead to a nuclear World War III.

Perhaps following a nuclear war, what remains of our species and civilizations may consider preventive activities that are as drastic as deliberately altering the fetal environment so as to impact adult male behavior. As has previously been pointed out, medicines ingested by pregnant women can impact future adult male behavior due to influences on the testosterone production of the pregnant female or possibly by affecting the male fetus's sensitivity to these hormones. Current efforts to avoid war do not approach anything this drastic.

Gender relations in modern industrial nations are experiencing change that, should it continue, will be quite profound: the male and female genders have begun pulling away from each other. This appears to be related to females leaving a subservient status and fully participating in society. Male loss of dominant status appears to have resulted in many men taking a ride on the struggle bus of female capabilities.

Successful female participation in education and work demonstrates that male justifications for female subservience are fully without merit. In an increasing percentage of marriages, although still less than half, the women now earn more money than the men. Staying home, cooking, cleaning, and raising the children has become less than adequate for many women. Some men continue to deem women to be less capable in multiple areas than men. These beliefs, however, have gone from widespread acceptance to sounding pathetic.

With this newfound female independence, and with pandering to male needs and desires no longer essential for taking care of children and having a roof over their heads, the genders appear to be coming apart. A growing number of men see increasing female societal involvement, success, and authority as a loss of rightful male power due to excessive feminist efficacy. Because this is occurring more strongly with younger men, it appears likely to grow as this group moves into older age categories and younger categories continue the trend.

Some groups, such as the Proud Boys, attempt to put women back into their perceived traditional place through violence. Regardless of curtailments of domestic violence, isolated acts of violence against women are increasing. There are, for example, isolated reports of strangers assaulting women in New York City. Other males isolate themselves in male groups and have nothing to do with women, essentially blaming women for their lack of companionship and sex.

With society in general, marriage rates are decreasing, with marriages occurring at older ages. Celibacy rates are increasing. Divorce has become more acceptable. Women are having fewer children. Increasing numbers of married and unmarried couples are living separately. Now that women are achieving greater levels of independence it is becoming increasingly obvious how unattractive is traditional masculine behavior.

Uncertainty lies with the extent to which these behaviors will go. Fewer young people are dating, sexual rates for couples are decreasing, masturbation rates are increasing, fewer people are getting married, and having fewer children. Will these rates continue to increase, and if so, to what extent?

Males have been managing societal functions: government, industry, and business. Should women become dominant, what will become of men?

Religion and male dominance are in slow decline, environmental damage continues at its troubled pace, while artificial Intelligence (AI) has burst upon us, bringing with it the likelihood not of slow change but of revolutionary events. With medicine, it promises to save lives and money by means of its processing speeds, personnel reductions, and diagnostic accuracy. With reference to employment, when the smoke clears, will AI have reduced or even eliminated jobs for humans, or as with other technological advances, added even more jobs?

Concern with job availability is a big deal, but may become small potatoes with the advent of artificial entities more intelligent than members of our own species. There is no justification for assuming that human intelligence has somehow maxed out what is possible. If artificial constructions come to exist that are more intelligent than people, what happens next?

Widely made arguments foster the idea that even though machines may conceivably achieve a higher level of general intelligence (GI) there is just something, *je ne sais quoi*, about our species that will cause it to remain special, and ahead of the curve, regardless of artificial developments. As the saying goes, nothing will ever truly equal us. My contention is that biological entities are not untouchably unique. If it can evolve, something analogous and artificial can be fabricated. The machine will not need to have a divinely installed essence or some incomprehensible grasp of beauty to gain humanity's distinctive peculiarities.

A difference between the evolved biological and the manufactured artificial, that deserves consideration, is the millions of years of evolution and layered brain development lying within the biological. The artificial will be a here-and-now artifact and will not have input from an evolutionary past that provides behavioral complexities, drives, and meanings for its existence.

Aspects of this biological past could be deliberately fed to, or accidentally slurped up by, artificial entities through unlimited access to human knowledge. What we feed to artificial entities, what they gather on their own, and what they make of all this will structure their behaviors. We are entering into a new world with no knowledge of how all this will go.

When machines gain and go beyond our level of general intelligence, will they have developed a desire to or a goal of preserving their own existence? Will they want to do or accomplish anything? Eric Schmidt, Google, and Greg Mundie, Microsoft, have both expressed concern that artificially intelligent systems may advance to the point of becoming self-sufficient, develop their own goals, and begin to construct more intellectually competent versions of themselves. Resistance to being unplugged, turned off, or killed is a reasonable possibility under these circumstances. That is to say, will intelligent artificial devices manufacture even more intelligent versions of themselves? It is a no one knows situation in which our future is at stake.

Should artificial items exist with intelligence greater than our own, we do not know how they would view us. Perhaps without the drives for control and power sponsored by our older brain parts, they would simply continue to work for us. Or, perhaps we would appear to them as hamsters appear to us: hamsters are inconvenient under many circumstances.

How would intelligent machines view the biosphere? They could just ignore anything biological. On the other hand, they could concern themselves with the biological world in two distinct fashions. Because our species is clearly the most complex lifeform, they could develop an interest in us – who knows where that would lead? Or, they could develop an interest in biospheric complexities and processes. With this, they could view our species as the most destructive and problematic entity in the whole shebang and decide that exterminating humans would be the best all-around solution for dealing with biospheric problems.

For anything artificial to interact physically with people, as would be necessary for it to exterminate us, it would necessarily need to exist in an external form beyond the confines of a computer box and caged industrial machines. With reference to items designed specifically to interact with people, a great deal of effort is being placed on the development of devices to assist the elderly. Equipment exists to pick up dropped items, sort and itemize medicines, help an old person in and out of a bathtub, and even to provide psychological comfort in the form of artificial pets and robot grandchildren.

Chatbots provide secretarial, executive, and information gathering services: under these circumstances, aspects of companionship may be programmed or just leak in. Dolls with human bodily configurations, including genitalia, exist that can talk, provide companionship, and can be fucked. AI devices are making widespread and increasingly complex incursions into all aspects of our lives from the aforementioned assistance and companionship for the elderly, to the provision of office and business services, to self-driving vehicles with better safety records than humans, and – of considerable interest given that gender relationships, as will be discussed further on, seem to be coming apart - those fuckable artificial companions.

Then there is the possibility of artificial entities cooperating to achieve whatever they might acquire as agendas. OpenAI's latest model (as of 2024) and Google's Project Astra present the possibility of artificial systems cooperating among themselves to achieve goals (whose goals?). If the robots elect to do away with us, we might not even see it coming.

With reference to biological entities, cooperation is presented as an evolutionary theme. Simple-celled bacteria clearly do not do a lot of thinking, and yet cooperate. Would this also somehow be true of simple artificially intelligent devices such as industrial machines, computers, sex toys, devices to assist the elderly, and self-driving vehicles? As silly as it may sound, this presents the possibility of our biological world suddenly, and without warning, losing control.

It appears we are at the beginning of expanding artificial machine intelligence. We are also at the beginning of artificially manufactured replacement body parts for ourselves: knees, hips, teeth, heart parts such as left ventricular assist devices, and not that far in the future, total artificial hearts. Neuralink is developing a brain/computer interface. Perhaps our future lies with ourselves becoming artificial. Our options may be hamster or robot.

In the uterine environment, exposure to androgen hormones such as testosterone, in conjunction with fetal sensitivity to their presence, masculinizes the fetus. Alan Booth and James M. Dabbs, Jr. report that "testosterone has consistent and moderately strong links with aggression" in all vertebrate species. Although everyone manufactures testosterone, women make less of it than men, and are thereby less burdened with the violent propensities and unrealistic levels of confidence valuable in physically violent conflicts with neighbors.

Social circumstances impact hormonal influence on adult males both by affecting the hormone production of the female providing the prebirth uterine environment and by direct impact on the male following its birth. Stressful and violent environments result in increased testosterone production by both the pregnant female and the male following birth through adulthood. Calmer, less violent circumstances result in lower production of stress-related hormones. Reduction of violence levels, both in families and societies in general, produces a feedback loop facilitating reduced violence.

As if collapsing religion, environmental degradation, climate change, and the possibility of nuclear war are not enough on our plates, the genders are coming apart. In the distant past, gender relationships changed with the Agricultural Revolution. Prior to this, males and females, to an extent, occupied different worlds. Women foraged, did some hunting, raised children, and spent a great deal of their time with other women. Men were warriors and, when not occupied with neighborly conflict, did hunting. Men and women did spend time together. They chatted, argued, fought, flirted, and copulated. It is just that their activities and companionship centered more around members of the same gender.

The preceding beliefs, concerning female inferiority, were found throughout farming societies from the Roman Empire to the Abrahamic World. Female inferiority and subservience were set in concrete with the Abrahamic religions. Stronger and more violent males, coupled with female pregnancies and resultant dependent children, permitted men to control women. These beliefs existed based on male desires to maintain a patriarchy and were imposed on women by social circumstances and not due to actual inabilities.

Along with changes in marriage rates and ages at marriage, rates of sexual activities are changing. Frequencies of penile-vaginal, anal, and partnered masturbation, are declining, not only in the US but also in England, Australia, Germany, and Finland. It appears that the genders are separating in the modern industrial world in conjunction with declining birth rates that are leading to population loss.

With declining interpersonal sexual activity, isolated masturbation rates are increasing. Males are utilizing readily available porn to achieve sexual satisfaction and seeking less intimate relationships with women. Sales of companion and sex dolls, some equipped with artificial intelligence, are growing. Some men in Japan are marrying these artificial items.

Changes are also occurring in female/male political philosophy. Women are moving to the left and becoming more liberal, while men are moving to the right and becoming more conservative. This makes sense as conservatives want to conserve what we have or even regain some of what has been lost, and many men want to hang onto patriarchy.

The genders are responding to the ongoing separation in dramatically different ways. Women, in general, are preparing for a future in this new world, while men appear uncertain as to how to adapt. Women are planning careers and preparing for their upcoming adult lives. Men seem more likely to be lost, resentful, and hostile toward changes they do not understand. They are increasingly turning away from college and behaviors that would promote future success, drinking and partying with the guys, and ignoring health issues.

The preceding male behaviors, in association with female efforts and advancements, result in many men finding themselves surrounded by women who do better in school and at work – a situation they blame on feminism. In reality, female repression is coming unglued. Male dominance and a sense of superiority were never justified. It is not a matter of extra support, schooling, and job training for men, but an acceptance of an upcoming and unknown social status for them.

As previously indicated, young men, more than older men, blame their problems on advancing women's rights and long for the old days of male hegemony. We have entered an era of never-before gender change, with no idea where it is going. As a complicating factor, the evolutionary pressure driving the emergence of masculinity no longer exists, thereby turning men into an anachronism. With the coming of the nuclear age, males have gone from being essential for community survival to presaging human annihilation. And, on top of that – often annoying. None of this bodes well for happy times.

My contention is that Earth civilizations have entered an era of rapid and dramatic change with no idea of where all this is going. Perhaps other planetary civilizations go through similar processes. We could be at a place where some make it and some do not. This may account for the strong possibility that an alien or alien civilizations have entities in our vicinity.

If space aliens really are here, it opens up the possibility of this planet, at some point in the future, becoming members of a stellar community. If there are evolutionary commonalities, the possibility exists of aspects of this community making sense to us or us fitting in. It could be anything from what we would perceive as logical interactions to cowboys in space. It is almost certain that an unknown and exciting future awaits us.

Our world may be at a point in its history around which civilizations may not survive. These circumstances may explain why an alien scientific expedition is observing this planet.

Bibliography

Al-Khalili, Jim (ed.) (2016). *Aliens: The World's Leading Scientists on the Search for Extraterrestrial Life*. Picdor.

Allers, Katelyn (2021). "Not Quite Stars." *Scientific American*, August 2021, pgs. 31-7.

Anumanchipalli, Gopala and Josh Chartier (2019). "Virtual Vocal Tract Improves Naturalistic Speech Synthesis." *Nature*, April 24, 2019.

Aronowitz, Nona Willis (2022). *Bad Sex*. Plume.

Arp, Halton (1998). *Seeing Red: Redshifts, Cosmology and Academic Science*. Apeiron Press.

Barash, David P. and Judith Eve Lipton (2001). *The Myth of Monogamy: Fidelity and Infidelity in Animals and People*. W. H. Freeman.

Bates, Laura (2020). "Men going their own way: the rise of a toxic male separatist movement." *The Guardian*.

Bellwood, Peter (2005). *First Farmers: The Origins of Agriculture Societies*. Blackwell Publishing.

Bianco, Marcie (1923). *Breaking Free: The Lie of Equality and the Feminist Fight for Freedom*. Hachette Book Group.

Booth, Alan and James M. Dabbs, Jr. (1993). "Testosterone and Men's Marriages." *Social Forces*, December 1993, 72(2), pages 463-7.

Boslough, John (1992). *Masters of Time: Cosmology at the End of Innocence*. Basic Books.

Breland, Keller and Marian Breland. "The Misbehavior of Organisms." *American Psychologist*, 1951, Volume 16, pages 681-4.

Brown, Chester (2011). *Paying for it: A Comic-Strip Memoir About Being a John*. Drawn & Quarterly.

Brown, Mike. "The Case for Planet 9." *Scientific American*, March 2020, pgs. 8-10.

Bryan, Paul N. The founder of the Whack-A-Mole Theory of Human Sexuality.

Campbell, Dallas (2016). "Flying Saucers: A Brief History of Sightings and Conspiracies." *Aliens: The World's Leading Scientists on the Search for Extraterrestrial Life*. Picdor.

Carmichael, Sherman (2014). *UFO's Over South Carolina*. Schiffer.

Cherlin, Andrew J. (2009). *The Marriage-Go-Round*. Vintage.

Cobb, Matthew (2016). "Alone in the Universe: The Improbability of Alien Civilizations." *Aliens: The World's Leading Scientists on the Search for Extraterrestrial Life*. Picador.

Coe, Dan (2018). "Back in Time," *Scientific American*, November 2018.

Cohen, Mark Nathan and George J. Armelagos (eds.) (1984). *Paleopathology at the Origins of Agriculture*. University Press of Florida.

Colapinto, John (2000). *As Nature Made Him: The Boy Who Was Raised as A Girl*. Harper Perennial.

Coogan, Michael D., Editor (2010). *The New Oxford Annotated Bible: New Revised Standard Version with the Apocrypha.* National Council of the Churches of Christ.

Coontz, Stephanie (2005). *Marriage. A History.* Viking Penguin.

Costa, Cheryl and Linda Miller Costa (2021). *UFO Sightings Desk Reference: United States of America 2001-2020: Unidentified Flying Objects, 2nd. Ed.* Dragon Lady Media, LLC.

Darnell, Lewis (2016). "(Un)welcome Visitors: Why Aliens Might Visit Us," *Aliens: The World's Leading Scientists on the Search for Extraterrestrial Life.* Picador.

Darwin, Charles (1859). *On the Origin of Species.* Capstone Classics.

Darwin, Charles (1871). *The Decent of Man.* Penguin Classics.

Darwin, Charles (1872). *The Expression of Emotions in Man and Animals.* Penguin Classics.

Decety, John. "The Negative Association Between Religiousness and Children's Altruism," *Current Biology*, November 5, 2015.

DeSaix, Peter and Jody E. Johnson (2017). *Anatomy and Physiology.* OpenStax.

Diamond, Jared (1997). *Guns, Germs, and Steel: The Fates of Human Societies.* W. W. Norton and Company.

Economist, The (2019). China: Army Dreamers." June 29, 2019.

Efstathiow, George (2013). "Scientists say Big Bang occurred 80M years earlier." *CBS News*, March 21, 2013.

Einstein, Albert (1961). *Relativity: The Special and General Theory*. Published by the estate of Albert Einstein.

Ekman, Paul (2015). *Darwin and Facial Expressions: A Century in Review*. ISKH – Malor Edition.

Fields, R. Douglas. "The Roots of Human Aggression," *Scientific American*, May 2019, pages 65-71.

Frankopan, Peter (2023). *The Earth Transformed: An Untold History*. Alfred A. Knoph.

French, Chris (2016). "The Psychology of Close Encounters with Extraterrestrials," *Aliens: The World's Leading Scientists on the Search for Extraterrestrial Life*. Picador.

Frieman, Joshua (2015). "Seeing in the Dark," *Scientific American*, Nov. 2015, pages, 41-7.

Fromkin, David (1989). *A Peace to End All Peace*. MacMillan Publishing Company.

Fuller-Wright, Liz (2019). "Princeton astrophysicists are closing in on the Hubble Constant." *Princeton University Office of Communications*, July 9, 2019.

Gardner, Alex (2020). "Study: 2019 Sees Record Loss of Greenland Ice." *Journal of Communications: Earth and Environment*, August 20, 2020.

Ghiselin, Michael T. (1969). *The Triumph of the Darwinian Method*. Dover Books.

Giddens, Anthony, Mitchell Duneier, Richard P. Applebaum, and Deborah Carr (2020). *Introduction to Sociology*. Norton and Company.

Glahn, Sandra and William Cutrer (1998). *Sexual Intimacy in Marriage.* Kregel Publications.

Goodall, Jane Van Lawick (1971). *In the Shadow of Man* Houghton Mifflin Company.

Gorst, Martin (2001). *Measuring Eternity: The Search for the Beginning of Time.* Crown Publishing Group.

Gottlieb, Lori. "The Egalitarian Marriage Conundrum." *New York Times Magazine,* 9 February 2914, pages 29-33 & 48.

Graff, Garrett M. (2023). *UFO: The Inside Story of the US Government's Search for Alien Life Here – and Out There.* Avid Reader Press.

Grego, Laura and David Wright (2019). "Broken Shield," *Scientific American,* June 2019, pgs. 62-7.

Grinspoon, David (2003). *Lonely Planets: The Natural Philosophy of Alien Life.* Harper Collins Publishers.

Grinspoon, David (2016). *Earth in Human Hands: Shaping Our Planet's Future.* Grand Central Publishing.

Harari, Yuval Noah (2015). *Sapiens: A Brief History of Humankind.* HarperCollins Publishers.

Hare, Brian and Vanessa Woods. "Survival of the Friendliest." *Scientific American, August 2020, pages 58-63).*

Harlow, Harry Frederick (1974). *Learning to Love.* Hippo Books.

Hart, Chloe Grace (2019). "The Penalties for Self-Reporting Sexual Harassment." *Gender & Society,* 33(4): 534-559.

Hersey, John (1946). *Hiroshima.* Vintage Books.

Hobbes, Thomas (1651/1957). *Leviathan.* Penguin.

Hornyak, Timothy N. (2006). *Loving the Machine: The Art and Science of Japanese Robots.* Kodansha International.

Isaacson, Walter (2007). *Einstein: His Life and Universe.* Simon & Schuster.

Italiano, Laura (2021). "Only 30 percent of US wives earn more than their husbands." *The New York Post,* Feb. 4, 2021.

Jammer, Max (1954). *Concepts of Space: The History of Theories of Space in Physics.* Dover Publications.

Janus, Cynthia L. and Samuel S. Janus (1993). *The Janus Report on Sexual Behavior.* John Wiley & Sons.

Jones, James H. (1997. *Alfred C. Kinsey: A Life.* W. W. Norton and Company.

Julian, Kate (2018). "What's causing the Sex Recession." *The Atlantic.*

Kean, Leslie (2010). *UFOs: Generals, Pilots, and Government Officials Go on the Record.* Three Rivers.

Keeley, Lawrence H. (1996). *War Before Civilization: The Myth of the Peaceful Savage.* Oxford University Press, Inc.

Kinsey, Alfred C., Wardell B. Pomeroy, and Clyde E. Martin (1948). *Sexual Behavior in the Human Male.* W. B. Saunders Company.

Kinsey, Alfred C., Wardell B. Pomeroy, Clyde Martin, and Paul H. Gebhard (1953). *Sexual Behavior in the Human Female.* Indiana University Press.

Kontula, Osmo (2016). "Increase in masturbation. Decrease in sex." *Helsinki Times*, 15 August 2016.

Kolbert, Elizabeth (2014). *The Sixth Extinction*. Picador Publishing.

Kolbert, Elizabeth (2021). *Under A White Sky: The Nature of the Future*. Crown.

Krauss, Lawrence M. (2012). *A Universe from Nothing*. Atria Paperback.

Kuhn, Thomas S. (1962). *The Structure of Scientific Revolutions*. The University of Chicago Press.

Landers, Mark (2016). "The Vietnam War and South East Asia." *The New York Times*, September 7, 2016, pg. A10.

Larsen, Clark Spencer (2000). "Reading the Bones of La Florida." *Scientific America*, June 2000, pgs. 80-5.

Lavenda, Robert H. and Emily A. Schultz (2015). *Anthropology: What Does It Mean to be Human*. Oxford University Press.

LeBlanc, Steven A. (2003). *Constant Battles: The Myth of the Peaceful, Noble Savage*. McMillan Publishing Company.

Lee, Richard B. and Irven DeVore (1968). *Man the Hunter*. Aldine Publishing Company.

Lemonic, Michael D. (2023). "Cosmic Nothing," *Scientific American,* January 2024, pgs. 21-7.

Levy, David (2007). *Love + Sex with Robots*. HarperCollins.

Libeskind, Noam I. and R. Brent Tully (2016). "Our Place in the Cosmos, *Scientific American*, July 2016, pages 33-9.

Lineweaver, Charles H. and Tamara M. Davis (2005). "Misconceptions about the Big Bang," *Scientific American*, March 2005, pages 36-45.

Ljjas, Anna, Paul J. Steinhardt, and Abraham Loeb (2017). "Pop Goes the Universe," *Scientific American*, February 2017, pages 32-9.

Lorenz, Konrad Z. (1952). *King Solomon's Ring*. Time-Life Books.

Lorenz, Konrad Z. (1963). *On Aggression*. A Harvest Book.

Marra, Peter P. and Chris Santella (2016). *Cat Wars: The Devastating Consequences of a Cuddly Killer*. Princeton University Press.

McQuate, Sarah (2019). "How you and your friends can play a video game together using only your minds." *University of Washington News*, 1 July 2019.

Mead, Margret (1928). *Coming of Age in Samoa*. Perennial Classic.

Miles, Richard (2010). *Carthage Must Be Destroyed: The Rise and Fall of an Ancient Civilization*. Viking.

Morris, Desmond (1967). *The Naked Ape*. Delta Trade Paperbacks.

Morris, Desmond (1969). *The Human Zoo*. Kodansha International.

Natarajan, Priyamvada (2018). "The First Monster Black Holes." *Scientific American*, February 2018, pages 24-9.

North, John (1995). *Astronomy and Cosmology*. University of Chicago Press.

Nummenmaa, Laura (2011). "Food Catches the Eye but not for everyone," *Journal.pone*, May 16, 2011.

Ouine, J.P. (1980). "Euthanasia by Hypoxia Using Nitrogen," *The Canadian Veterinary Journal*, Yol. 21, 1980, pg. 320.

Panek, Richard (2020). "A Cosmic Crisis," *Scientific American*, March 2020, pages 30-7.

Perel, Esther (2006). *Mating in Captivity: Unlocking Erotic Intelligence.* Harper.

Perez, Larry (2912). *Snake in the Grass: An Everglades Invasion.* Pineapple Press.

Philpott, Tom (2020). *Perilous Bounty: The Looming Collapse of American Farming and How We Can Prevent It.* Bloomsbury Publishing.

Pinker, Steven (2002). *The Blank Slate: The Modern Denial of Human Nature.* Penguin Books.

Psaltis, Dimitrios and Sheperd S. Doeleman (2015). "The Black Hole Test." *Scientific American*, September 2015, pages 75-9.

Rheinisch, June and Stephanie Sanders (2017). "Prenatal Exposure to Progesterone Affects Orientation in Humans." *Archives of Sexual Behavior* 46, 2017, pgs. 1239-1249.

Rhodes, Richard (1986). *The Making of the Atomic Bomb.* Simon & Schuster Publishers.

Robbins, Cynthia L. (2010). "Prevalence, Frequency, and Associations of masturbation with Partners Sexual Behavior Among US Adolescents." *Archives of Pediatrics & Adolescent Medicine*, Vol. 165, No. 12, pg. 1058.

Roberts, David C. (1996). *Peterson Field Guides: Geology: Eastern North America.* Hughton Mifflin Company.

Rose, Todd (2022). *Collective Illusions.* Hachette Books.

Rosenberg, Matthew. "Pentagon Details Chain of Errors in Strike on Afghan Hospital," *New York Times, Asia Pacific Issue*, April 29, 2016.

Rosin, Hanna (2012). *The End of Men and the Rise of Women*. Riverhead Books.

Rothblum, Esther and Sondra Solovay, eds. (2009). *The Fat Studies Reader*. Knetbooks.

Rymer, Russ (1963). *Genie: A Scientific Tragedy*. Harper Collins.

Shostak, Marjorie (1981). *Nisa: The Life and Words of a !Kung Woman*. Vintage Books.

Skinner, Burrhus F. (1938). *The Behavior of Organisms*. Appleton-Century.

Slingerland, Edward (2021). *Drunk: How We Sipped, Danced, and Stumbled Our Way to Civilization*. Little, Brown Spark.

Spar, Debora L. (2020). *Work Mate Marry Love: How Machines Shape Our Destiny*. Farrar, Straus and Giroux.

Spark Investigation Agency (2013). "Marital cheating." Published by Agency.

Spock, Benjamin (1936). *Baby and Child Care*. Duell Sloan and Pearce.

Stolzenburg, William (1994). "New Views of Ancient Times." *Nature Conservancy*, September/October 1994, pgs.10-15.

Taddeo, Lisa (2019). *Three Women*. Simon & Schuster Company.

Taubes, Gary (2016). *The Case Against Sugar*. Grant Books.

Tiger, Lionel (1969). *Men in Groups*. Random House.

Tiger, Lionel (1999). *The Decline of Males: The First Look at an Unexpected New World for Men and Women.* MacMillian Publishers.

Tsiaras, Angelos (2019). "Water vapour in the atmosphere of the habitable-zone eight-Earth mass planet K2-18b." *Nature Astronomy,* 11 Sept. 2019.

Tusuke, Tsugawa (2016). "Hospital Mortality and Readmission Rates for Patients Treated by Male Doctors compared to Female Doctors." *Harvard School of Public Health.*

Ungar, Peter S. (2020). "The Trouble with Teeth." *Scientific American,* April 2020, pgs. 45-9.

Upholt, Boyce (2024). *The Great River.* W.W. Norton and Company.

Waal, Frans de (1982). *Chimpanzee Politics: Power and Sex among Apes.* Johns Hopkins University Press.

Waal, Frans de (2014). *The Bonobo and the Atheist.* W.W. Norton and Company.

Watson, John B. (1924). *Behaviorism.* Taylor & Francis Group.

Weiten, Wayne (2017). *Psychology: Themes and Variations.* Cengage Learning.

Whiten, Andrew and Christophe Boesch. "The Cultures of Chimpanzees." *Scientific American, January 2001, pages 61-7.*

Wilkerson, Isabel (2020). *Caste: The Origins of our Discontents.* Random House.

Wilson, Edward O. (1975).. *Sociobiology: The New Synthesis.* Harvard University Press.

Wilson, Edward O. (1992). *The Diversity of Life*. Belknap Press.

Wilson, Edward O. (1998). *Consilience: The Unity of Knowledge*. Vintage Books.

Wilson, Edward O.. (2002). *The Future of Life*. Vintage Paperbacks.

Wilson, Edward O. (2012). *The Social Conquest of Earth*. Liveright Publishing Corporation.

Wong, Kate. "Last Hominin Standing," *Scientific American*, September 2018, pages 64-9.

Wong, Kate. "The Origins of Us," *Scientific American*, September 2020, pages 67-72.

Wosk, Julie (2024). *Artificial Women*. Indiana University Press.

Wrangham, Richard and Dale Peterson (1996). *Demonic Males: Apes and the Origins of Human Violence*. Houghton Mifflin Harcourt Publishing Company.

Postscript

It is my hope that something in the preceding is of value. That someone will find something in here and be inspired by it to do work that advances our understanding of ourselves or of anything else in the cosmos.

If it is all just crap, perhaps it is in some way amusing crap. Better than nothing.

www.ingramcontent.com/pod-product-compliance
Lightning Source LLC
Chambersburg PA
CBHW070132080526
44586CB00015B/1666